# Use R!

*Series Editors:*
Robert Gentleman   Kurt Hornik   Giovanni Parmigiani

For further volumes:
http://www.springer.com/series/6991

Jim Albert • Maria Rizzo

# R by Example

 Springer

Jim Albert
Department of Mathematics and Statistics
Bowling Green State University
Bowling Green Ohio
USA
albert@bgsu.edu

Maria Rizzo
Department of Mathematics and Statistics
Bowling Green State University
Bowling Green Ohio
USA
mrizzo@bgsu.edu

*Series Editors:*
Robert Gentleman
Program in Computational Biology
Division of Public Health Sciences
Fred Hutchinson Cancer Research Center
1100 Fairview Ave. N. M2-B876
Seattle, Washington 98109-1024
USA

Kurt Hornik
Department für Statistik und Mathematik
Wirtschaftsuniversität Wien Augasse 2-6
A-1090 Wien
Austria

Giovanni Parmigiani
The Sidney Kimmel Comprehensive
Cancer Center at Johns Hopkins University
550 North Broadway
Baltimore, MD 21205-2011
USA

ISBN 978-1-4614-1364-6          ISBN 978-1-4614-1365-3 (eBook)
DOI 10.1007/978-1-4614-1365-3
Springer New York Dordrecht Heidelberg London

Library of Congress Control Number: 2011941603

Springer is part of Springer Science+Business Media (www.springer.com)

# Preface

*R by Example* is an example-based introduction to the R [40] statistical computing environment that does not assume any previous familiarity with R or other software packages. R is a statistical computing environment for statistical computation and graphics, and it is a computer language designed for typical and possibly very specialized statistical and graphical applications. The software is available for unix/linux, Windows, and Macintosh platforms under general public license, and the program is available to download from www.r-project.org. Thousands of contributed packages are also available as well as utilities for easy installation.

The purpose of this book is to illustrate a range of statistical and probability computations using R for people who are learning, teaching, or using statistics. Specifically, this book is written for users who have covered at least the equivalent of (or are currently studying) undergraduate level calculus-based courses in statistics. These users are learning or applying exploratory and inferential methods for analyzing data and this book is intended to be a useful resource for learning how to implement these procedures in R.

Chapters 1 and 2 provide a general introduction to the R system and provide an overview of the capabilities of R to perform basic numerical and graphical summaries of data. Chapters 3, 4, and 5 describe R functions for working with categorical data, producing statistical graphics, and implementing the exploratory data analysis methods of John Tukey. Chapter 6 presents R procedures for basic inference about proportions and means. Chapters 7 through 10 describe the use of R for popular statistical models, such as regression, analysis of variance (ANOVA), randomized block designs, two-way ANOVA, and randomization tests. The last section of the book describes the use of R in Monte Carlo simulation experiments (Chapter 11), Bayesian modeling (Chapter 12), and Markov Chain Monte Carlo (MCMC) algorithms to simulate from probability distributions (Chapter 13).

One general feature of our presentation is that R functions are presented in the context of interesting applications with real data. Features of the useful R function lm, for example, are best communicated through a good regression example. We have tried to reflect good statistical practice through the examples that we present in all chapters. An undergraduate student should easily be able to relate our R work on, say, regression with the regression material that is taught in his statistics course. In each chapter, we include exercises that give the reader practice in implementing the R functions that are discussed.

The data files used in the examples are available in R or provided on our web site. A few of the data files can be input directly from a web page, and there are also a few that found in the recommended packages (installed with R) or contributed packages (installed by the user when needed). The web page for this book is `personal.bgsu.edu/~mrizzo/Rx`.

Remarks or tips about R are identified by the symbol **R$_x$** to set them apart from the main text. In the examples, R code and output appears in bold monospaced type. Code that would be typed by the user is identified by the leading prompt symbol >. Scripts for some of the functions in the examples are provided in files available from the book web site; these functions are shown without the prompt character.

The R manuals and examples in the help files use the arrow assignment operators <- and ->. However, in this book we have used the equal sign = operator for assignment, rather than <-, as novice users may find it easier to type the = symbol.

R functions and keywords are collected at the beginning of the Index. Examples are also indexed; see the entry 'Example' in the Index.

Bowling Green, Ohio                                              *Jim Albert*
                                                                 *Maria Rizzo*

# Contents

# Notation and Abbreviations

| | |
|---|---|
| $\in$ | is an element of |
| $\propto$ | is proportional to |
| $\Gamma(a)$ | complete gamma function, $\Gamma(a) = \int_0^\infty t^{a-1} \exp(-t)\, dt$ |

| | |
|---|---|
| cdf | cumulative distribution function |
| csv | comma separated values (file format) |
| E | expected value |
| GUI | graphical user interface |
| iid | independent and identically distributed |
| IQR | interquartile range |
| log | natural logarithm (base $e$) |
| NID | normally distributed and independent |
| QQ | quantile-quantile (plot) |
| MC | Monte Carlo (methods) |
| MCMC | Markov chain Monte Carlo |
| M-H | Metropolis-Hastings (algorithm) |
| MSE | mean squared error |
| MST | mean square for treatments |
| SS.total | total sum of squares |
| SSE | sum of squares for error (residual sum of squares) |
| SST | sum of squares for treatments |

# Chapter 1
# Introduction

R is a statistical computing environment. It is free (open source) software for statistical computation and graphics [40] and a computer language designed for typical statistical and graphical applications. The R distribution includes the ability to save and run commands stored in script files, and an integrated editor in the R Graphical User Interface (R-GUI). It is available for most platforms including unix/linux, PC, and Macintosh platforms. Thousands of contributed packages are available, and users are provided tools to make packages.

At the core of R is an interpreted computer language. This language provides the logical control of branching and looping, and modular programming using functions. The base R distribution contains functions and data to implement and illustrate most common statistical procedures, including regression and ANOVA, classical parametric and nonparametric tests, cluster analysis, density estimation, and much more. An extensive suite of probability distribution functions and generators are provided, as well as a graphical environment for exploratory data analysis and creating presentation graphics.

On the history and evolution of R, see the R-FAQ [26] and resources on the R home page at http://www.R-project.org/.

## 1.1 Getting Started

R is an interpreted language; that is, the system processes commands entered by the user, who types the commands at the command prompt, or submits the commands from a file called a script. We assume that our readers use R at a graphics workstation running a windowing system, such as Windows, Macintosh, or X window systems. In a window system, users interact with R through the R console. Except for the simplest operations, most users will prefer to type commands in a script (see Section 1.1.3) to save retyping and to separate commands from results. However, let us begin by working directly at the command prompt.

When we use the command line interface, each command or expression to be evaluated is typed at the command prompt, and immediately evaluated when the Enter key is pressed at the end of a syntactically complete statement. It is helpful to remember the following tips.

- Press the up-arrow key to recall commands and edit them.
- Use the Esc (Escape) key to cancel a command.

### 1.1.1 Preliminaries

Remarks or tips about R are identified by the symbol $\mathbf{R_x}$ to set them apart from the main text.

$\mathbf{R_x}$ **1.1** *The right-to-left assignment operators are the left arrow* <- *and equal sign* =. *For example, borrowing a line from Example 1.3, either method below*

```
> x = c(109, 65, 22, 3, 1)
> x <- c(109, 65, 22, 3, 1)
```

*creates the vector* $(109, 65, 22, 3, 1)$ *and assigns it to x. Borrowing another line from Example 1.3, either method below*

```
> y = rpois(200, lambda=.61)
> y <- rpois(200, lambda=.61)
```

*assigns the result of the* **rpois** *function to y. Notice that the equal sign inside the parentheses is not an assignment operator; it passes the value of an argument (lambda) to the function* **rpois**.

The R manuals and examples in the help files use the arrow assignment operators <- and ->. However, in this book we have used the equal sign = operator for assignment, rather than <-, as novice users may find it easier to type the = symbol.

In the examples, R code and output appears in bold monospaced type as in the remark $\mathbf{R_x}$ 1.1 above. Code that would be typed interactively by the user or submitted from an R script is identified by the leading prompt symbol >. Scripts for some of the functions in the examples are provided in files available from the book web site; these functions are shown in the book without the prompt character.

Data files and scripts used in the examples are available on our web site at personal.bgsu.edu/~mrizzo/Rx. Some data files can be downloaded directly from a connection to a url.

## 1.1.2 Basic operations

Some basic operations with vectors are illustrated in the following example. The R commands are entered at the prompt in the R console window. The prompt character is > and when a line is continued the prompt changes to +. (The prompt symbols can be changed.)

*Example 1.1 (Temperature data).* Average annual temperatures in New Haven, CT, were recorded in degrees Fahrenheit, as

```
Year                 1968   1969   1970  1971
Mean temperature     51.9   51.8   51.9    53
```

(This data is part of a larger data set in R called *nhtemp*.) The combine function c creates a vector from its arguments, and the result can be stored in user-defined vectors. We use the combine function to enter our data and store it in an object named temps.

```
> temps = c(51.9, 51.8, 51.9, 53)
```

To display the value of temps, one simply types the name.

```
> temps
[1] 51.9 51.8 51.9 53.0
```

Suppose that we want to convert the Fahrenheit temperatures (F) to Celsius temperatures (C). The formula for the conversion is $C = \frac{5}{9}(F - 32)$. It is easy to apply this formula to all of the temperatures in one step, because arithmetic operations in R are *vectorized*; operations are applied element by element. For example, to subtract 32 from every element of temp, we use

```
> temps - 32
[1] 19.9 19.8 19.9 21.0
```

Then (5/9)*(temps - 32) multiplies each difference by 5/9. The temperatures in degrees Celsius are

```
> (5/9) * (temps - 32)
[1] 11.05556 11.00000 11.05556 11.66667
```

In 1968 through 1971, the mean annual temperatures (Fahrenheit) in the state of Connecticut were 48, 48.2, 48, 48.7, according to the National Climatic Center Data web page. We store the state temperatures in CT, and compare the local New Haven temperatures with the state averages. For example, one can compute the annual differences in mean temperatures. Here CT and temps are both vectors of length four and the subtraction operation is applied element by element. The result is the vector of four differences.

```
> CT = c(48, 48.2, 48, 48.7)
> temps - CT
[1] 3.9 3.6 3.9 4.3
```

The four values in the result are differences in mean temperatures for 1968 through 1971. It appears that on average New Haven enjoyed slightly warmer temperatures than the state of Connecticut in this period.

*Example 1.2 (President's heights).* An article in Wikipedia [54] reports data on the heights of Presidents of the United States and the heights of their opponents in the presidential election. It has been observed [53, 48] that the taller presidential candidate typically wins the election. In this example, we explore the data corresponding to the elections in the television era. In Table 1.1 are the heights of the presidents and their opponents in the U.S. presidential elections of 1948 through 2008, extracted from the Wikipedia article.

**Table 1.1** Height of the election winner in the Electoral College and height of the main opponent in the U.S. Presidential elections of 1948 through 2008.

| Year | Winner | Height | | Opponent | Height | |
|------|--------|--------|--|----------|--------|--|
| 2008 | Barack Obama | 6 ft 1 in | 185 cm | John McCain | 5 ft 9 in | 175 cm |
| 2004 | George W. Bush | 5 ft 11.5 in | 182 cm | John Kerry | 6 ft 4 in | 193 cm |
| 2000 | George W. Bush | 5 ft 11.5 in | 182 cm | Al Gore | 6 ft 1 in | 185 cm |
| 1996 | Bill Clinton | 6 ft 2 in | 188 cm | Bob Dole | 6 ft 1.5 in | 187 cm |
| 1992 | Bill Clinton | 6 ft 2 in | 188 cm | George H.W. Bush | 6 ft 2 in | 188 cm |
| 1988 | George H.W. Bush | 6 ft 2 in | 188 cm | Michael Dukakis | 5 ft 8 in | 173 cm |
| 1984 | Ronald Reagan | 6 ft 1 in | 185 cm | Walter Mondale | 5 ft 11 in | 180 cm |
| 1980 | Ronald Reagan | 6 ft 1 in | 185 cm | Jimmy Carter | 5 ft 9.5 in | 177 cm |
| 1976 | Jimmy Carter | 5 ft 9.5 in | 177 cm | Gerald Ford | 6 ft 0 in | 183 cm |
| 1972 | Richard Nixon | 5 ft 11.5 in | 182 cm | George McGovern | 6 ft 1 in | 185 cm |
| 1968 | Richard Nixon | 5 ft 11.5 in | 182 cm | Hubert Humphrey | 5 ft 11 in | 180 cm |
| 1964 | Lyndon B. Johnson | 6 ft 4 in | 193 cm | Barry Goldwater | 5 ft 11 in | 180 cm |
| 1960 | John F. Kennedy | 6 ft 0 in | 183 cm | Richard Nixon | 5 ft 11.5 in | 182 cm |
| 1956 | Dwight D. Eisenhower | 5 ft 10.5 in | 179 cm | Adlai Stevenson | 5 ft 10 in | 178 cm |
| 1952 | Dwight D. Eisenhower | 5 ft 10.5 in | 179 cm | Adlai Stevenson | 5 ft 10 in | 178 cm |
| 1948 | Harry S. Truman | 5 ft 9 in | 175 cm | Thomas Dewey | 5 ft 8 in | 173 cm |

Section 1.5 illustrates several methods for importing data from a file. In this example we enter the data interactively as follows. The continuation character + indicates that the R command is continued.

```
> winner = c(185, 182, 182, 188, 188, 188, 185, 185, 177,
+    182, 182, 193, 183, 179, 179, 175)
> opponent = c(175, 193, 185, 187, 188, 173, 180, 177, 183,
+    185, 180, 180, 182, 178, 178, 173)
```

(Another method for entering data interactively is to use the scan function. See Example 3.1 on page 79.) Now the newly created objects winner and opponent are each vectors of length 16.

```
> length(winner)
[1] 16
```

The year of the election is a regular sequence, which we can generate using the sequence function `seq`. Our first data value corresponds to year 2008, so the sequence can be created by

```
> year = seq(from=2008, to=1948, by=-4)
```

or equivalently by

```
> year = seq(2008, 1948, -4)
```

According to the Washington Post blog [53], Wikipedia misstates "Bill Clinton's height, which was measured during official medical exams at 6 foot-2-1/2, making him just a tad taller than George H.W. Bush." We can correct the height measurement for Bill Clinton by assigning a height of 189 cm to the fourth and fifth entries of the vector `winner`.

```
> winner[4] = 189
> winner[5] = 189
```

The sequence operator : allows us to perform this operation in one step:

```
> winner[4:5] = 189
```

The revised values of `winner` are

```
> winner
 [1] 185 182 182 189 189 188 185 185 177 182 182 193 183 179 179 175
```

Are presidents taller than average adult males? According to the National Center for Health Statistics, in 2005 the average height for an adult male in the United States is 5 feet 9.2 inches or 175.768 cm. The sample mean is computed by the `mean` function.

```
> mean(winner)
[1] 183.4375
```

Interestingly, the opponents also tend to be taller than average.

```
> mean(opponent)
[1] 181.0625
```

Next, we use vectorized operations to compute the differences in the height of the winner and the main opponent, and store the result in `difference`.

```
> difference = winner - opponent
```

An easy way to display our data is as a data frame:

```
> data.frame(year, winner, opponent, difference)
```

The result is displayed in Table 1.2. Data frames are discussed in detail in Section 1.4.

We see that most, but not all, of the differences in height are positive, indicating that the taller candidate won the election. Another approach to determining whether the taller candidate won is to compare the heights with the logical operator >. Like the basic arithmetic operations, this operation is vectorized. The result will be a vector of logical values (`TRUE`/`FALSE`) having the same length as the two vectors being compared.

**Table 1.2** Data for Example 1.2.

```
> data.frame(year, winner, opponent, difference)
   year winner opponent difference
1  2008    185      175          10
2  2004    182      193         -11
3  2000    182      185          -3
4  1996    189      187           2
5  1992    189      188           1
6  1988    188      173          15
7  1984    185      180           5
8  1980    185      177           8
9  1976    177      183          -6
10 1972    182      185          -3
11 1968    182      180           2
12 1964    193      180          13
13 1960    183      182           1
14 1956    179      178           1
15 1952    179      178           1
16 1948    175      173           2
```

```
> taller.won = winner > opponent
> taller.won
 [1]  TRUE FALSE FALSE  TRUE  TRUE  TRUE  TRUE  TRUE FALSE
[10] FALSE  TRUE  TRUE  TRUE  TRUE  TRUE  TRUE
```

On the second line, the prefix [10] indicates that the output continues with the tenth element of the vector.

The `table` function summarizes discrete data such as the result in the vector `taller.won`.

```
> table(taller.won)
taller.won
FALSE  TRUE
    4    12
```

We can use the result of `table` to display percentages if we divide the result by 16 and multiply that result by 100.

```
> table(taller.won) / 16 * 100
taller.won
FALSE  TRUE
   25    75
```

Thus, in the last 16 elections, the odds in favor of the taller candidate winning the election are 3 to 1.

Several types of graphs of this data may be interesting to help visualize any pattern. For example, we could display a barplot of differences using the `barplot` function. For the plot we use the `rev` function to reverse the order of the differences so that the election year is increasing from left to right. We also provide a descriptive label for both axes.

```
> barplot(rev(difference), xlab="Election years 1948 to 2008",
+    ylab="Height difference in cm")
```

The barplot of differences in heights is shown in Figure 1.1.

It would also be interesting to display a scatterplot of the data. A scatterplot of loser's heights vs winner's height for election years 1798 through 2004 appears in the Wikipedia article [54]. A simple version of the scatterplot (not shown here) can be obtained in R by

```
> plot(winner, opponent)
```

Chapter 4 "Presentation Graphics" illustrates many options for creating a custom graphic such as the scatterplot from the Wikipedia article.

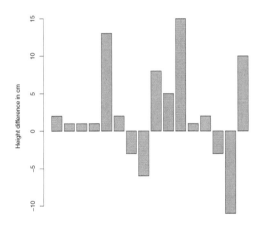

Election years 1948 to 2008

**Fig. 1.1** Barplot of the difference in height of the election winner in the Electoral College over the height of the main opponent in the U.S. Presidential elections. Height differences in centimeters for election years 1948 through 2008 are shown from left to right. The electoral vote determines the outcome of the election. In 12 out of these 16 elections, the taller candidate won the electoral vote. In 2000, the taller candidate (Al Gore) did not win the electoral vote, but received more popular votes.

*Example 1.3 (horsekicks).* This data set appears in several books; see e.g. Larsen and Marx [30, p. 287]. In the late 19th century, Prussian officers collected data on deaths of soldiers in 10 calvary corps recording fatalities due to horsekicks over a 20 year period. The 200 values are summarized in Table 1.3.

To enter this data, we use the *combine* function c.

**Table 1.3** Fatalities due to horsekick for Prussian calvary in Example 1.3

| Number of deaths, $k$ | Number of corps-years in which $k$ fatalities occurred |
|:---:|:---:|
| 0 | 109 |
| 1 | 65 |
| 2 | 22 |
| 3 | 3 |
| 4 | 1 |
| | 200 |

```
> k = c(0, 1, 2, 3, 4)
> x = c(109, 65, 22, 3, 1)
```

To display a bar plot of the frequencies, we use the `barplot` function. The function `barplot(x)` produces a barplot like Figure 1.2, but without the labels below the bars. The argument `names.arg` is optional; it assigns labels to display below the bars. Figure 1.2 is obtained by.

```
> barplot(x, names.arg=k)
```

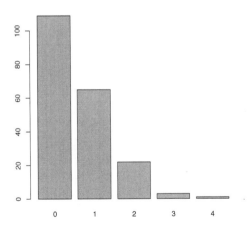

**Fig. 1.2** Frequency distribution for Prussian horsekick data in Example 1.3.

The relative frequency distribution of the observed data in x is easily computed using vectorized arithmetic in R. For example, the sample proportion of 1's is $65/200 = 0.545$. The expression `x/sum(x)` divides every element of

the vector x by the sum of the vector (200). The result is a vector the same
length as x containing the sample proportions of the death counts 0 to 4.

```
> p = x / sum(x)
> p
[1] 0.545 0.325 0.110 0.015 0.005
```

The center of this distribution can be estimated by its sample mean, which
is

$$\frac{1}{200}\sum_{i=1}^{200} x_i = \frac{109(0)+65(1)+22(2)+3(3)+1(4)}{200}.$$

$$= 0.545(0)+0.325(1)+0.110(2)+0.015(3)+0.005(4).$$

The last line is simply the sum of p*k, because R computes this product
element by element ("vectorized"). Now we can write the sample mean formula
as the sum of the vector p*k. The value of the sample mean is then assigned
to r.

```
> r = sum(p * k)
> r
[1] 0.61
```

Similarly, one can compute an estimate of the variance. Apply the computing
formula for variance of a sample $y_1, \ldots, y_n$:

$$s^2 = \frac{1}{n-1}\sum_{i=1}^{n}(y_i - \bar{y})^2.$$

Here the sample mean is the value r computed above and

$$s^2 = \frac{1}{n-1}\left\{109(0-r)^2+65(1-r)^2+22(2-r)^2+3(3-r)^2+1(4-r)^2\right\},$$

so the expression inside the braces can be coded as x*(k-r)^2. The sample
variance v is:

```
> v = sum(x * (k - r)^2) / 199
> v
[1] 0.6109548
```

Among the counting distributions that might fit this data (binomial, geometric, negative binomial, Poisson, etc.) the Poisson is the one that has
equal mean and variance. The sample mean 0.61 and sample variance 0.611
are almost equal, which suggests fitting a Poisson distribution to the data.
The Poisson model has probability mass function

$$f(k) = \frac{\lambda^k e^{-\lambda}}{k!}, \quad k \geq 0, \tag{1.1}$$

where $\lambda = \sum_{k=0}^{\infty} k f(k)$ is the mean of the distribution. The sample mean 0.61 is our estimate of the population mean $\lambda$. Substituting the sample mean for $\lambda$ in the density (1.1), the corresponding Poisson probabilities are

```
> f = r^k * exp(- r) / factorial(k)
> f
[1] 0.5433509 0.3314440 0.1010904 0.0205551 0.0031346
```

R has probability functions for many distributions, including Poisson. The R density functions begin with "d" and the Poisson density function is dpois. The probabilities above can also be computed as

```
> f = dpois(k, r)
> f
[1] 0.5433509 0.3314440 0.1010904 0.0205551 0.0031346
```

**R$_x$ 1.2** *R provides functions for the density, cumulative distribution function (CDF), percentiles, and for generating random variates for many commonly applied distributions. For the Poisson distribution these functions are* **dpois**, **ppois**, **qpois**, *and* **rpois**, *respectively. For the normal distribution these functions are* **dnorm**, **pnorm**, **qnorm**, *and* **rnorm**.

How well does the Poisson model fit the horsekick data? In a sample of size 200, the expected counts are $200f(k)$. Truncating the fraction using floor we have

```
>  floor(200*f)   #expected counts
[1] 108   66   20    4    0
>  x              #observed counts
[1] 109   65   22    3    1
```

for $k = 0, 1, 2, 3, 4$, respectively. The expected and observed counts are in close agreement, so the Poisson model appears to be a good one for this data.

One can alternately compare the Poisson probabilities (stored in vector f) with the sample proportions (stored in vector p). To summarize our comparison of the probabilities in a matrix we can use rbind or cbind. Both functions bind vectors together to form matrices; with rbind the vectors become rows, and with cbind the vectors become columns. Here we use cbind to construct a matrix with columns k, p, and f.

```
> cbind(k, p, f)
      k     p          f
[1,]  0 0.545 0.5433509
[2,]  1 0.325 0.3314440
[3,]  2 0.110 0.1010904
[4,]  3 0.015 0.0205551
[5,]  4 0.005 0.0031346
```

It appears that the observed proportions p are close to the Poisson(0.61) probabilities in f.

## 1.1.3 R Scripts

Example 1.3 contains several lines of code that would be tedious to retype if one wants to continue the data analysis. If the commands are placed in a file, called an R script, then the commands can be run using source or copy-paste. Using the source function causes R to accept input from the named source, such as a file.

Open a new R script for editing. In the R GUI users can open a new script window through the *File* menu. Type the following lines of "horsekicks.R" (below) in the script. It is a good idea to insert a few comments. Comments begin with a # symbol.

Using the source function, auto-printing of expressions does not happen. We added print statements to the script so that the values of objects will be printed.

```
──────────────────── horsekicks.R ────────────────────
# Prussian horsekick data
k = c(0, 1, 2, 3, 4)
x = c(109, 65, 22, 3, 1)
p = x / sum(x)          #relative frequencies
print(p)

r = sum(k * p)    #mean
v = sum(x * (k - r)^2) / 199   #variance
print(r)
print(v)
f = dpois(k, r)
print(cbind(k, p, f))
```

At this point it is convenient to create a working directory for the R scripts and data files that will be used in this book. To display the current working directory, type getwd(). For example, one may create a directory at the root, say /Rx. Then change the working directory through the File menu or by the function setwd, substituting the path to your working directory in the quotation marks below. On our system this has the following effect.

```
> getwd()
[1] "C:/R/R-2.13.0/bin/i386"
> setwd("c:/Rx")
> getwd()
[1] "c:/Rx"
```

Save the script as "horsekicks.R" in your working directory. Now the file can be sourced by the command

```
source("horsekicks.R")
```

and all of the commands in the file will be executed.

**R$_x$ 1.3** *Unlike* Matlab *.m files, an R script can contain any number of functions and commands.* Matlab *users may be familiar with defining a function* *by writing an .m file, where each .m file is limited to exactly one function. Function syntax is covered in Section 1.2.*

**R$_\mathbf{x}$ 1.4** *Here are a few helpful shortcuts for running part of a script.*

- *Select lines and click the button 'Run line or selection' on the toolbar.*
- *Copy the lines, and then paste the lines at the command prompt.*
- *(For Windows users:) To execute one or more lines of the file in the R GUI editor, select the lines and type Ctrl-R.*
- *(For Macintosh users:) One can execute lines of a file by selecting the lines and typing Command-Return.*

*Example 1.4 (Simulated horsekick data).* For comparison with Example 1.3, in this example we use the random Poisson generator **rpois** to simulate 200 random observations from a Poisson($\lambda = 0.61$) distribution. We then compute the relative frequency distribution for this sample. Because these are randomly generated counts, each time the code below is executed we obtain a different sample and therefore the results of readers will vary slightly from what follows.

```
> y = rpois(200, lambda=.61)
> kicks = table(y)     #table of sample frequencies
> kicks
y
  0   1   2   3
105  67  26   2
> kicks / 200          #sample proportions
y
    0     1     2     3
0.525 0.335 0.130 0.010
```

Comparing this data with the theoretical Poisson frequencies:

```
> Theoretical = dpois(0:3, lambda=.61)
> Sample = kicks / 200
> cbind(Theoretical, Sample)
  Theoretical Sample
0  0.54335087  0.525
1  0.33144403  0.335
2  0.10109043  0.130
3  0.02055505  0.010
```

The computation of mean and variance is simpler here than in Example 1.3 because we have the raw, ungrouped data in the vector **y**.

```
> mean(y)
[1] 0.625
> var(y)
[1] 0.5571608
```

It is interesting that the observed Prussian horsekicks data seems to fit the Poisson model better than our simulated Poisson($\lambda = 0.61$) sample.

## *1.1.4 The R Help System*

The R Graphical User Interface has a Help menu to find and display online documentation for R objects, methods, data sets, and functions. Through the Help menu one can find several manuals in PDF form, an html help page, and help search utilities. The help search utility functions are also available at the command line, using the functions `help` and `help.search`, and the corresponding shortcuts ? and ??. These functions are described below.

- `help("keyword")` displays help for "keyword".
- `help.search("keyword")` searches for all objects containing "keyword".

The quotes are usually optional in `help`, but would be required for special characters such as in `help("[")`. Quotes are required for `help.search`. When searching for help topics, keep in mind that R is case-sensitive: for example, `t` and `T` are different objects.

One or two question marks in front of a search term also search for help topics.

- `?keyword` (short for `help(keyword)`).
- `??keyword` (short for `help.search("keyword")`).

Try entering the following commands to see their effect.

```
?barplot        #searches for barplot topic
??plot          #anything containing "plot"

help(dpois)         #search for "dpois"
help.search("test") #anything containing "test"
```

The last command above displays a list including a large number of statistical tests implemented in the R.

One of the features of R online help is that most of the keywords documented include examples appearing at the end of the page. Users can try one or more of the examples by selecting the code and then copy-paste to the console. R also provides a function `example` that runs all of the examples if any exist for the keyword. To see the examples for the function **mean**, type **example(mean)**. The examples are then executed and displayed at the console with a special prompt symbol (**mean>**) that is specific to the keyword.

```
> example(mean)

mean> x <- c(0:10, 50)

mean> xm <- mean(x)

mean> c(xm, mean(x, trim = 0.10))
[1] 8.75 5.50
```

```
mean> mean(USArrests, trim = 0.2)
  Murder  Assault UrbanPop     Rape
    7.42   167.60    66.20    20.16
>
```

For many of the graphics functions, the documentation includes interesting examples. Try example(curve) for an overview of what the curve function can do. The system will prompt the user for input as it displays each graph.

A glossary of R functions is available online in "Appendix D: Function and Variable Index" of the manual "Introduction to R" [49], and the "R Reference Manual" [41] has a comprehensive index by function and concept. These manuals are included with the R distribution, and also available online on the R project home page[1] at the line "Manuals" under "Documentation".

## 1.2 Functions

The R language allows for modular programming using functions. R users interact with the software primarily through functions. We have seen several examples of functions above. In this section, we discuss how to create user-defined functions.

The syntax of a function is

```
f = function(x, ...) {
  }
```

or

```
f <- function(x, ...) {
  }
```

where f is the name of the function, x is the name of the first argument (there can be several arguments), and ... indicates possible additional arguments. Functions can be defined with no arguments, also. The curly brackets enclose the body of the function. The return value of a function is the value of the last expression evaluated.

*Example 1.5 (function definition).* R has a function var that computes the unbiased estimate of variance, usually denoted by $s^2$. Occasionally, one requires the maximum likelihood estimator (MLE) of variance,

$$\hat{\sigma}^2 = \frac{1}{n}\sum_{i=1}^{n}(x_i - \overline{x})^2 = \frac{n-1}{n}s^2.$$

A function to compute $\hat{\sigma}^2$ can be created as follows.

---

[1] www.r-project.org

```
var.n = function(x) {
  v = var(x)
  n = NROW(x)
  v * (n - 1) / n
}
```

The `NROW` function computes the number of observations in x. The value v *
(n-1)/n evaluated on the last line is returned. Note: it would also be correct
(but unnecessary) to replace the last line of the function var.n with

```
    return(v * (n - 1) / n)
```

Before this user-defined function can be used, one must input the code
so that the function, in this case var.n, is an object in the R workspace.
Normally, one places functions in a script file and uses the source function
(or copy and paste to the command line) to submit them. Here is an example
that computes $s^2$ and $\hat{\sigma}^2$ for the temperature data of Example 1.1.

```
> temps = c(51.9, 51.8, 51.9, 53)
> var(temps)
[1] 0.3233333
> var.n(temps)
[1] 0.2425
```

*Example 1.6 (functions as arguments).* Many of the available R functions require functions as arguments. An example is the `integrate` function, which
implements numerical integration; one must supply the integrand as an argument. Suppose that we need to compute the beta function, which is defined
as

$$B(a,b) = \int_0^1 x^{a-1}(1-x)^{b-1}\,dx,$$

for constants $a > 0$ and $b > 0$. First we write a function that returns the
integrand evaluated at a given point $x$. The additional arguments $a$ and $b$
specify the exponents.

```
f = function(x, a=1, b=1)
    x^(a-1) * (1-x)^(b-1)
```

The curly brackets are not needed here because there is only one line in the
function body. Also, we defined default values $a = 1$ and $b = 1$, so that if $a$ or
$b$ are not specified, the default values will be used. The function can be used
to evaluate the integrand along a sequence of $x$ values.

```
> x = seq(0, 1, .2)   #sequence from 0 to 1 with steps of .2
> f(x, a=2, b=2)
[1] 0.00 0.16 0.24 0.24 0.16 0.00
```

This *vectorized* behavior is necessary for the function argument of the in-
tegrate function; the function that evaluates the integrand must accept a
vector as its first argument and return a vector of the same length.

Now the numerical integration result for $a = b = 2$ can be obtained by

```
> integrate(f, lower=0, upper=1, a=2, b=2)
0.1666667 with absolute error < 1.9e-15
```

Actually, R provides a function `beta` to compute this integral. We can compare our numerical integration result to the value returned by the `beta` function:

```
> beta(2, 2)
[1] 0.1666667
```

See `?Special` for more details on the beta and other special functions of mathematics.

$\mathbf{R_x}$ *1.5 The* `integrate` *function is an example of a function syntax with extra arguments (...). The complete function syntax is*

```
integrate(f, lower, upper, ..., subdivisions=100,
          rel.tol = .Machine$double.eps^0.25, abs.tol = rel.tol,
          stop.on.error = TRUE, keep.xy = FALSE, aux = NULL)
```

*The three dots are additional arguments to be passed to the integrand function* $f$. *In our example, these extra arguments were* $a$ *and* $b$.

*Example 1.7 (graphing a function using* **curve**). R provides the `curve` function to display the graph of a function. For example, suppose that we wish to graph the function

$$f(x) = x^{a-1}(1-x)^{b-1}$$

for $a = b = 2$, which is the integrand in Example 1.6. This is easily obtained as

```
> curve(x*(1-x), from=0, to=1, ylab="f(x)")
```

See Figure 1.3 for the result. The function argument in `curve` is always written as a function of `x`. The optional argument `ylab` specifies the label for the vertical axis.

## 1.3 Vectors and Matrices

A *vector* in R contains a finite sequence of values of a single type, such as a sequence of numbers or a sequence of characters. A *matrix* in R is a two dimensional array of values of a single type.

Common operations on vectors and matrices are illustrated with the following probability example. A more detailed introduction to vectors and matrices in R is provided in the Appendix.

*Example 1.8 (Class mobility).* The following model of class mobility is discussed in Ross [42, Example 4.19, p. 207]. Assume that the class of a child (lower, middle, or upper class) depends only on the class of his/her parents.

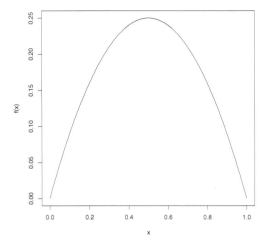

**Fig. 1.3** Plot of the beta function for parameters $a = 2, b = 2$ in Example 1.7.

The class of the parents is indicated by the row label. The entries in the table below correspond to the chance that the child will transition to the class indicated by the column label.

```
       lower middle upper
lower  0.45  0.48   0.07
middle 0.05  0.70   0.25
upper  0.01  0.50   0.49
```

To create a matrix with these transition probabilities, we use the `matrix` function. First, the vector of probabilities `probs` is constructed to supply the entries of the matrix. Then the matrix is defined by its entries, number of rows, and number of columns.

```
> probs = c(.45, .05, .01, .48, .70, .50, .07, .25, .49)
> P = matrix(probs, nrow=3, ncol=3)
> P
     [,1] [,2] [,3]
[1,] 0.45 0.48 0.07
[2,] 0.05 0.70 0.25
[3,] 0.01 0.50 0.49
```

Notice that the values are entered by column; to enter the data by row, use the optional argument `byrow=TRUE` in the `matrix` function. Matrices can optionally have row names and column names. In this case, row names and column names are identical, so we can assign both using

```
> rownames(P) <- colnames(P) <- c("lower", "middle", "upper")
```

and the updated value of P is

```
> P
       lower middle upper
lower    0.45   0.48  0.07
middle   0.05   0.70  0.25
upper    0.01   0.50  0.49
```

In the matrix $P = (p_{ij})$, the probability $p_{ij}$ in the $i$-th row and $j$-th column is the probability of a transition from class $i$ to class $j$ in one generation.

   This type of matrix has rows that sum to 1 (because each row is a probability distribution on the three classes). This fact can be verified by the rowSums function.

```
> rowSums(P)
 lower middle  upper
     1      1      1
```

Another approach uses the apply function. It requires specifying the name of the matrix, MARGIN (row=1, column=2), and FUN (function) as its arguments.

```
> apply(P, MARGIN=1, FUN=sum)
lower middle  upper
    1      1      1
```

   It can be shown that the transition probabilities for two generations are given by the product $P^2 = PP$, which can be computed by the matrix multiplication operator %*%.

```
> P2 = P %*% P
> P2
          lower middle  upper
lower    0.2272 0.5870 0.1858
middle   0.0600 0.6390 0.3010
upper    0.0344 0.5998 0.3658
```

$R_X$ 1.6 *Here we did not use the syntax* P^2 *because* P^2 *squares every element of the matrix and the result is the matrix* $(p_{ij}^2)$, *not the matrix product.*

   To extract elements from the matrix, the [row, column] syntax is used. If the row or column is omitted, this specifies all rows (columns). In two generations, the probability that descendants of lower class parents can transition to upper class is in row 1, column 3:

```
> P2[1, 3]
[1] 0.1858
```

and the probability distribution for lower class to (lower, middle, upper) is given by row 1:

```
> P2[1, ]
 lower middle  upper
0.2272 0.5870 0.1858
```

   After several generations, each row of the transition matrix will be approximately equal, with probabilities $p = (l, m, u)$ corresponding to the percentages of lower, middle, and upper class occupations. After eight transitions, the probabilities are P8:

```
> P4 = P2 %*% P2
> P8 = P4 %*% P4

> P8
             lower    middle      upper
lower   0.06350395 0.6233444 0.3131516
middle  0.06239010 0.6234412 0.3141687
upper   0.06216410 0.6234574 0.3143785
```

It can be shown that the limiting probabilities are 0.07, 0.62, and 0.31. For
the solution $p$, see Ross [42, p. 207].

$\mathbf{R_X}$ **1.7** *To enter a matrix of constants, as in this example, it is usually easier
to enter data by rows using **byrow=TRUE** in the **matrix** function. Compare the
following with the example on page 17.*

```
> Q = matrix(c(  0.45,  0.48,  0.07,
+                0.05,  0.70,  0.25,
+                0.01,  0.50,  0.49), nrow=3, ncol=3, byrow=TRUE)
> Q
     [,1] [,2] [,3]
[1,] 0.45 0.48 0.07
[2,] 0.05 0.70 0.25
[3,] 0.01 0.50 0.49
```

*The "by row" format makes it easier to see (visually) the data vector as a
matrix in the code.*

$\mathbf{R_X}$ **1.8** *Matrix operations for numeric matrices in R:*

1. *Elementwise multiplication:* *
   *If matrices $A = (a_{ij})$ and $B = (b_{ij})$ have the same dimension, then* A*B *is
   evaluated as the matrix with entries $(a_{ij}b_{ij})$.*
2. *If $A = (a_{ij})$ is a matrix then* A^r *is evaluated as the matrix with entries
   $(a_{ij}^r)$.*
3. *Matrix multiplication:* %*%
   *If $A = (a_{ij})$ is an $n \times k$ matrix and $B = (b_{ij})$ is a $k \times m$ matrix, then*
   A %*% B *is evaluated as the $n \times m$ matrix product $AB$.*
4. *Matrix inverse:*
   *If $A$ is a nonsingular matrix, the inverse of $A$ is returned by* solve(A).

   *For eigenvalues and matrix factorization, see* **eigen**, **qr**, **chol**, *and* **svd**.

## 1.4 Data Frames

Data frames are special types of objects in R designed for data sets that are
somewhat like matrices, but unlike matrices, the columns of a `data.frame`
can be different types such as numeric or character. Several data sets are
installed with R; a list of these data sets can be displayed by the command
`data()`. Most data sets that are provided with R are in data frame format.

## 1.4.1 Introduction to data frames

The data frame format is similar to a spreadsheet, with the variables corresponding to columns and the observations corresponding to rows. Variables in a data frame may be numeric (numbers) or categorical (characters or factors).

To get an initial overview of a data frame, we are usually interested in knowing the names of variables, type of data, sample size, numbers of missing observations, etc.

*Example 1.9 (USArrests).* The *USArrests* data records rates of violent crimes in the US. The statistics are given as arrests per 100,000 residents for assault, murder, and rape in each of the 50 US states in 1973. The percentage of the population living in urban areas is also given. Some basic functions to get started with a data frame are illustrated with this data set.

### Display all data

To simply display the data, type the name of the object, USArrests.

### Display top of data

To display the first few lines of data:

```
> head(USArrests)
           Murder Assault UrbanPop Rape
Alabama      13.2     236       58 21.2
Alaska       10.0     263       48 44.5
Arizona       8.1     294       80 31.0
Arkansas      8.8     190       50 19.5
California    9.0     276       91 40.6
Colorado      7.9     204       78 38.7
```

The result shows that we have four variables named Murder, Assault, UrbanPop, and Rape, and that the observations (rows) are labeled by the name of the state. We also see that the states appear to be listed in alphabetical order. All of the variables appear to be quantitative, which we expected from the description above.

### Sample size and dimension

How many observations are in this data set? (NROW, nrow, or dim)

```
> NROW(USArrests)
[1] 50
> dim(USArrests)  #dimension
[1] 50   4
```

The dimension (`dim`) of a data frame or a matrix returns a vector with the number of rows and number of columns. `NROW` returns the number of observations. We have 50 observations corresponding to the 50 states in the U.S.

## Names of variables

Get (or set) names of variables in the data frame:

```
> names(USArrests)
[1] "Murder"   "Assault"  "UrbanPop" "Rape"
```

## Structure of the data

Display information about the structure of the data frame (`str`):

```
> str(USArrests)
'data.frame':   50 obs. of  4 variables:
 $ Murder  : num  13.2 10 8.1 8.8 9 7.9 3.3 5.9 15.4 17.4 ...
 $ Assault : int  236 263 294 190 276 204 110 238 335 211 ...
 $ UrbanPop: int  58 48 80 50 91 78 77 72 80 60 ...
 $ Rape    : num  21.2 44.5 31 19.5 40.6 38.7 11.1 15.8 31.9 25.8 ...
```

The result of `str` gives the dimension as well as the name and type of each variable. We have two numeric type and two integer type variables. Although we can think of integer as a special case of numeric, they are stored differently in R.

**R$_x$ 1.9** *For many data sets, like **USArrests**, all of the data are numbers and in this case the data can be converted to a matrix using **as.matrix**. But in order to store the data in a matrix, all variables must be of the same type so R will convert the integers to numeric. Compare the result in matrix form:*

```
> arrests = as.matrix(USArrests)
> str(arrests)
 num [1:50, 1:4] 13.2 10 8.1 8.8 9 7.9 3.3 5.9 15.4 17.4 ...
 - attr(*, "dimnames")=List of 2
  ..$ : chr [1:50] "Alabama" "Alaska" "Arizona" "Arkansas" ...
  ..$ : chr [1:4] "Murder" "Assault" "UrbanPop" "Rape"
```

*This output shows that all of the data was converted to numeric, listed on the first line as **num**. The attributes (**attr**) are the row and column names (**dimnames**). The conversion preserved the row labels and converted the variable names to column labels. We used **names** to get the names of the variables in the data frame, but we would use **rownames**, **colnames** or **dimnames** (to get both) to get the row and/or column names. These last three functions can also be used on data frames.*

**Missing values**

The is.na function returns TRUE for a missing value and otherwise FALSE.
The expression is.na(USArrests) will return a data frame the same size as
USArrests where every entry is TRUE or FALSE. To quickly check if any of the
results are TRUE we use the any function.

```
> any(is.na(USArrests))
[1] FALSE
```

We see that USArrests does not contain missing values. A data set with
missing values is discussed in Example 2.3 on page 53.

## 1.4.2 Working with a data frame

In this section we illustrate some operations on data frames, and some basic
statistics and plots.

*Example 1.10 (USArrests, cont.).*

**Compute summary statistics**

Obtain appropriate summary statistics for each variable using summary. For
numeric data, the summary function computes a five-number summary and
sample mean.

```
> summary(USArrests)
     Murder          Assault         UrbanPop          Rape
 Min.   : 0.800   Min.   : 45.0   Min.   :32.00   Min.   : 7.30
 1st Qu.: 4.075   1st Qu.:109.0   1st Qu.:54.50   1st Qu.:15.07
 Median : 7.250   Median :159.0   Median :66.00   Median :20.10
 Mean   : 7.788   Mean   :170.8   Mean   :65.54   Mean   :21.23
 3rd Qu.:11.250   3rd Qu.:249.0   3rd Qu.:77.75   3rd Qu.:26.18
 Max.   :17.400   Max.   :337.0   Max.   :91.00   Max.   :46.00
```

If there were any missing values, the number of missing values would be
included in the summaries; see e.g. Example 2.3 on page 53. If any of our
variables were categorical, summary would tabulate the values for those variables.

From the summary it appears that the mean and median are approximately equal for all variables except Assault. The mean for Assault is larger
than the median, indicating that the assault data is positively skewed.

## Extract data from a data frame

The simplest way to extract data from a data frame uses the matrix-style [row, column] indexing.

```
> USArrests["California", "Murder"]
[1] 9
> USArrests["California", ]
           Murder Assault UrbanPop Rape
California      9     276       91 40.6
```

## Extract a variable using $

Variables can be extracted using the $ operator followed by the name of the variable.

```
> USArrests$Assault
 [1] 236 263 294 190 276 204 110 238 335 211  46 120 249 113
[15]  56 115 109 249  83 300 149 255  72 259 178 109 102 252
[29]  57 159 285 254 337  45 120 151 159 106 174 279  86 188
[43] 201 120  48 156 145  81  53 161
```

## Histograms

In the summary of the *USArrests* data frame, we observed that the distribution of assaults may be positively skewed because the sample mean is larger than the sample median. A histogram of the data helps to visualize the shape of the distribution. We show two versions of the histogram of Assault. The result of

```
> hist(USArrests$Assault)
```

is shown in Figure 1.4(a). The skewness is easier to observe in the second histogram (Figure 1.4(b)), obtained using

```
> library(MASS)      #need for truehist function
> truehist(USArrests$Assault)
```

There are obvious differences in the histograms; one is that the vertical scales differ. Another is that there are a different number of bins. Figure 1.4(a) is a frequency histogram, and Figure 1.4(b) is a probability histogram.

**R<sub>x</sub>** *1.10 Try the following command and compare the result with* **truehist** *in Figure 1.4(b):*

```
hist(USArrests$Assault, prob=TRUE, breaks="scott")
```

*The two histogram functions* **hist** *and* **truehist** *have different default methods for determining the bin width, and the* **truehist** *function by default produces a probability histogram. The optional arguments of* **hist** *above match the defaults of* **truehist**.

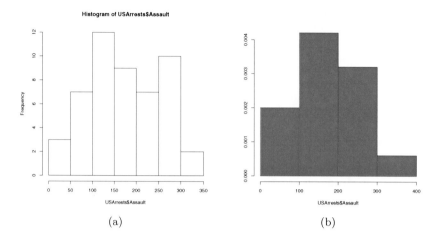

(a)                                                          (b)

**Fig. 1.4** Frequency histogram (a) using `hist` and probability histogram (b) using `truehist` for the assault data in `USArrests`.

### Attaching a data frame

If we `attach` the data frame, the variables can be referenced directly by name, without the dollar sign operator. For example, it is easy to compute the percent of crimes that are murders using vectorized operations.

```
> attach(USArrests)
> murder.pct = 100 * Murder / (Murder + Assault + Rape)
> head(murder.pct)
[1] 4.881657 3.149606 2.431702 4.031150 2.764128 3.152434
```

If a data frame is attached, it can be detached when it is no longer needed using the `detach` function.

An alternative to attaching a data frame is (sometimes) to use the `with` function. It is useful for displaying plots or summary statistics. However, variables created using `with` are local to it.

```
> with(USArrests, expr={
+    murder.pct = 100 * Murder / (Murder + Assault + Rape)
+    })
> murder.pct
Error: object 'murder.pct' not found
```

The documentation for `with` states that "assignments within expr take place in the constructed environment and not in the user's workspace." So all computations involving the variable `murder.pct` created in the `expr` block of the `with` function would have to be completed within the scope of the `expr` block. This can sometimes lead to unexpected errors in an analysis that can easily go undetected.

$\mathbf{R_{x}}$ **1.11** *Is it good programming practice to* **attach** *a data frame? A disadvantage is that there can be conflicts with names of variables already in the workspace. Many functions in R have a* **data** *argument, which allows the variables to be referenced by name within the list of arguments, without attaching the data frame. Unfortunately, some R functions (such as* **plot***) do not have a* **data** *argument. The use of* **with** *could lead to unexpected programming errors, especially for novices. Overall, we find that attaching a data frame is sometimes helpful to make code more readable.*

### Scatterplots and correlations

In the USArrests data all of the variables are numbers, so it is interesting to display scatterplots of different pairs of data to look for possible relations between the variables. We can display a single scatterplot using the plot function. Recall that above we have attached the data frame using attach so the variables can be referenced directly by name. To obtain a plot of murders vs percent urban population we use

```
> plot(UrbanPop, Murder)
```

and the result is shown in Figure 1.5. The plot does not reveal a very strong relation between murders and percent population. The pairs function can be used to display an array of scatterplots for each pair of variables.

```
> pairs(USArrests)
```

The pairs plot shown in Figure 1.6 conveys much information about the data. There appears to be a positive association between murder and assault rates, but weak or no association between murder and percent urban population. There is a positive association between rape and percent urban population.

The correlation statistic measures the degree of linear association between two variables. One can obtain correlation for a pair of variables or a table of correlation statistics for a group of variables using the cor function. By default it computes the Pearson correlation coefficient.

```
> cor(UrbanPop, Murder)
[1] 0.06957262
> cor(USArrests)
            Murder   Assault  UrbanPop      Rape
Murder   1.00000000 0.8018733 0.06957262 0.5635788
Assault  0.80187331 1.0000000 0.25887170 0.6652412
UrbanPop 0.06957262 0.2588717 1.00000000 0.4113412
Rape     0.56357883 0.6652412 0.41134124 1.0000000
```

All of the correlations are positive in sign. The small correlation of $r \doteq 0.07$ between Murder and UrbanPop is consistent with our interpretation of the scatterplot in Figure 1.5. There is a strong positive correlation ($r = 0.80$) between Murder and Assault, also consistent with the plot in Figure 1.6.

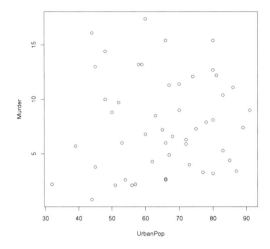

**Fig. 1.5** Scatterplot of murder rate vs percent urban population in USArrests.

## 1.5 Importing Data

A preliminary task for data analysis is to import data into R. Data sets may
be found on the web, in plain text (space delimited) files, spreadsheets, and
many other formats. R contains utility functions to import data in a variety
of formats. When a data set is imported into R, typically we store it in an
R data.frame object. See the examples of Section 1.4.

In this section we cover various methods of importing data, including:

- Entering data manually.
- Importing data from a plain text (ASCII) local file.
- Importing data given in tabular form on a web page.

### 1.5.1 Entering data manually

In the first few examples of this chapter we have seen how to enter data
vectors using the c (combine) function. The scan function is sometimes useful
for entering small data sets interactively. See Example 3.1 "Flipping a Coin"
in Chapter 3 for an example.

The concept for this book was originally inspired by students' questions in
a statistics course. After learning some R basics, students were eager to try
their textbook problems in R, but typically needed help to enter the data.

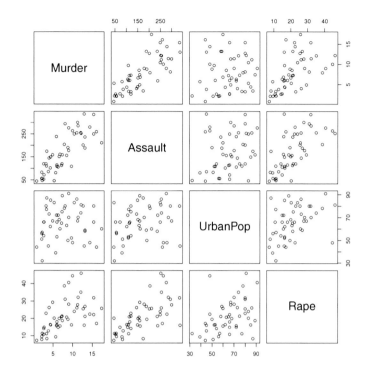

**Fig. 1.6** An array of scatterplots produced by `pairs` for the `USArrests` data.

The next example shows how to enter a very small data set that one might find among the exercises in a statistics textbook.

*Example 1.11 (Data from a textbook).* A table of gas mileages on four new models of Japanese luxury cars is given in Larsen & Marx [30, Question 12.1.1], which is shown in Table 1.4. The reader was asked to test if the four models give the same gas mileage, on average.

**Table 1.4** Gas mileage on four models of Japanese luxury cars, from a problem in Larsen & Marx [30, 12.1.1].

| Model | | | |
|---|---|---|---|
| A | B | C | D |
| 22 | 28 | 29 | 23 |
| 26 | 24 | 32 | 24 |
| | 29 | 28 | |

To solve the stated problem, we need to enter the data as two variables (gas mileage and model). One method of doing so is shown below. The `rep` (replicate) function is used to create the sequences of letters.

```
> y1 = c(22, 26)
> y2 = c(28, 24, 29)
> y3 = c(29, 32, 28)
> y4 = c(23, 24)
> y = c(y1, y2, y3, y4)
> Model = c(rep("A", 2), rep("B", 3), rep("C", 3), rep("D", 2))
```

We see that y and `Model` have been entered correctly:

```
> y
 [1] 22 26 28 24 29 29 32 28 23 24
> Model
 [1] "A" "A" "B" "B" "B" "C" "C" "C" "D" "D"
```

The data frame is created by

```
> mileages = data.frame(y, Model)
```

The character vector `Model` is converted to a factor by default when the data frame is created. We can check that the structure of our data frame is correct using the structure (`str`) function.

```
> str(mileages)
'data.frame':   10 obs. of  2 variables:
 $ y     : num  22 26 28 24 29 29 32 28 23 24
 $ Model: Factor w/ 4 levels "A","B","C","D": 1 1 2 2 2 3 3 3 4 4
```

and

```
> mileages
    y Model
1  22     A
2  26     A
3  28     B
4  24     B
5  29     B
6  29     C
7  32     C
8  28     C
9  23     D
10 24     D
```

Our data set is ready for analysis, which is left as an exercise for a later chapter (Exercise 8.1).

### 1.5.2 Importing data from a text file

It is often the case that data for analysis is contained in an external file (external to R) in plain text (ASCII) format. The data is typically delimited

or separated by a special character such as a space, tab, or comma. The
read.table function provides optional arguments such as sep for the sepa-
rator character and header to indicate whether or not the first row contains
variable names.

*Example 1.12 (Massachusetts Lunatics).* Importing data from a plain text
file can be illustrated with an example of a data set available on the website
"Data and Story Library" (DASL). The *Massachusetts lunatics* data is available at http://lib.stat.cmu.edu/DASL/Datafiles/lunaticsdat.html.

These data are from an 1854 survey conducted by the Massachusetts Commission on Lunacy. Fourteen counties were surveyed. The data can be copied
from the web page, and simply pasted into a plain text file. Although the
result is not nicely formatted, it is space delimited (columns separated by
spaces). The data is saved in "lunatics.txt" in our current working directory.
This data set should be imported into a data frame that has 14 rows and six
columns, corresponding to the following variables:

1. COUNTY = Name of county
2. NBR = Number of lunatics, by county
3. DIST = Distance to nearest mental health center
4. POP = County population , 1950 (thousands)
5. PDEN = County population density per square mile
6. PHOME = Percent of lunatics cared for at home

Use the read.table function to read the file into a data frame. The argument header=TRUE specifies that the first line contains variable names rather
than data. Type ?read.table for a description of other possible arguments.

```
> lunatics = read.table("lunatics.txt", header=TRUE)
```

The str (structure) function provides a quick check that 14 observations of
six variables were successfully imported.

```
> str(lunatics)
'data.frame':   14 obs. of  6 variables:
 $ COUNTY: Factor w/ 14 levels "BARNSTABLE","BERKSHIRE",..
 $ NBR   : int  119 84 94 105 351 357 377 458 241 158 ...
 $ DIST  : int  97 62 54 52 20 14 10 4 14 14 ...
 $ POP   : num  26.7 22.3 23.3 18.9 82.8 ...
 $ PDEN  : int  56 45 72 94 98 231 3252 3042 235 151 ...
 $ PHOME : int  77 81 75 69 64 47 47 6 49 60 ...
```

Now since lunatics is a relatively small data set, we can simply print it to
view the result of read.table that we used to import the data. Typing the
name of the data set causes it to be printed (displayed) at the console.

```
> lunatics
     COUNTY NBR DIST    POP PDEN PHOME
1  BERKSHIRE 119   97 26.656   56    77
2   FRANKLIN  84   62 22.260   45    81
3  HAMPSHIRE  94   54 23.312   72    75
4    HAMPDEN 105   52 18.900   94    69
```

```
5   WORCESTER 351   20  82.836   98   64
6   MIDDLESEX 357   14  66.759  231   47
7      ESSEX 377   10  95.004 3252   47
8    SUFFOLK 458    4 123.202 3042    6
9    NORFOLK 241   14  62.901  235   49
10   BRISTOL 158   14  29.704  151   60
11  PLYMOUTH 139   16  32.526   91   68
12 BARNSTABLE  78   44  16.692   93   76
13  NANTUCKET  12   77   1.740  179   25
14      DUKES  19   52   7.524   46   79
```

The *Massachusetts lunatics* data set is discussed in Example 7.8 of our regression chapter.

In the example above, the data in the file "lunatics.txt" is delimited by space characters. Often data is found in a spreadsheet format, delimited by tab characters or commas.

For tab-delimited files, simply change the **sep** argument to the tab character \t. An example appears in Chapter 5 on exploratory data analysis, where the tab-delimited data file "college.txt" is imported using the command

```
dat = read.table("college.txt", header=TRUE, sep="\t")
```

$\mathbf{R_x}$ **1.12** *The simplest way to import spreadsheet data is to save it in .csv format (comma separated values) or a tab-delimited format. The worksheet should contain only data and possibly the header with names. For .csv files, use the* **read.table** *function with* **sep=","** *as shown below*

```
> lunatics = read.table("lunatics.csv", header=TRUE, sep=",")
```

*or use* **read.csv** *for this type of file:*

```
> lunatics = read.csv("lunatics.csv")
```

### 1.5.3 Data available on the internet

Many of the interesting data sets that one may wish to analyze are available on a web page. R provides an easy way to access data from a file on the internet using the URL of the web page. The function **read.table** can be used to input data directly from the internet. This is illustrated in the following example.

*Example 1.13 (Digits of $\pi$).* The data file "PiDigits.dat" contains the first 5000 digits of the mathematical constant $\pi = 3.14159265358979323844\ldots$. The data is one of the *Statistical Reference Datasets* provided by the National Institute of Standards and Technology (NIST).[2] Documentation is inserted at the top of the file, and the digits start on line 61. We use the **read.table** function

---

[2] http://www.itl.nist.gov/div898/strd/univ/pidigits.html

with the complete URL[3] (web address) and `skip=60` to read the data starting at line 61. We display the first six digits to check the result:

```
pidigits = read.table(
  "http://www.itl.nist.gov/div898/strd/univ/data/PiDigits.dat",
  skip=60)
head(pidigits)

  V1
1 3
2 1
3 4
4 1
5 5
6 9
```

Here although we have only one variable, our data `pidigits` is a data frame. The data frame was automatically created because we used `read.table` to import the data and a default label of `V1` assigned to the single variable.

Are the digits of $\pi$ uniformly distributed? The digits can be summarized in a table by the `table` function and summarized graphically in a plot.

```
> table(pidigits)

pidigits
  0   1   2   3   4   5   6   7   8   9
466 531 496 461 508 525 513 488 491 521
```

For easier interpretation, it is more convenient to summarize proportions. We can convert the table to proportions in one step by dividing by 5000; this is another example of vectorized operations.

```
> prop = table(pidigits) / 5000    #proportions
> prop

pidigits
     0      1      2      3      4      5      6      7      8      9
0.0932 0.1062 0.0992 0.0922 0.1016 0.1050 0.1026 0.0976 0.0982 0.1042
```

Recall that the variance of a sample proportion is $p(1-p)/n$. If the true proportion is 0.1 for every digit, then the standard error (se) is

```
> sqrt(.1 * .9 / 5000)
[1] 0.004242641
```

However, if the true proportions are unknown, the sample estimates of proportions are used. In this case we obtain slightly different results for se. In the calculation of `se` the constants 0.1 and 0.9 are replaced by vectors of length 10, so the result is a vector of length 10 rather than a scalar. We can display the sample proportion plus or minus two standard errors using vectorized arithmetic. The `rbind` function is handy to collect the results together into a

---

[3] The URL should be enclosed in quotes and on a single line; otherwise an error message "cannot open the connection" occurs.

matrix for display. We round the result and include the estimate of standard error in the display.

```
> se.hat = sqrt(prop * (1-prop) / 5000)
> round(rbind(prop, se.hat, prop-2*se.hat, prop+2*se.hat), 4)
              0      1      2      3      4      5      6
prop     0.0932 0.1062 0.0992 0.0922 0.1016 0.1050 0.1026
se.hat   0.0041 0.0044 0.0042 0.0041 0.0043 0.0043 0.0043
         0.0850 0.0975 0.0907 0.0840 0.0931 0.0963 0.0940
         0.1014 0.1149 0.1077 0.1004 0.1101 0.1137 0.1112
              7      8      9
prop     0.0976 0.0982 0.1042
se.hat   0.0042 0.0042 0.0043
         0.0892 0.0898 0.0956
         0.1060 0.1066 0.1128
```

Here we see that none of the sample proportions falls outside of the interval $0.1 \pm 2\,\widehat{se}$.

A barplot helps to visualize the tabulated data. A horizontal reference line is added through 0.1 using `abline`. The plot is shown in Figure 1.7.

```
barplot(prop, xlab="digit", ylab="proportion")
abline(h = .1)
```

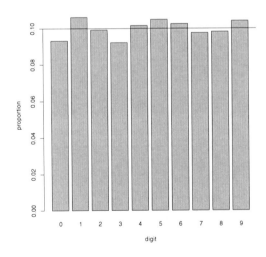

**Fig. 1.7** Barplot of proportion of digits in the mathematical constant $\pi$.

## 1.6 Packages

R functions are grouped into packages, such as the **base** package, **datasets**, **graphics**, or **stats**. A number of recommended packages are included in the R distribution; some examples are **boot** (bootstrap), MASS [50], and **lattice** (graphics). In addition, thousands of contributed packages are available to install. Type the command

```
library()
```

to display a list of installed packages. For example, on our system currently this command produces a list starting with

```
Packages in library 'C:/R/R-2.13.0/library':
```

```
base                     The R Base Package
boot                     Bootstrap R (S-Plus) Functions (Canty)
bootstrap                Functions for the Book "An Introduction to the
                         Bootstrap"
    ...
```

Each of these packages is currently installed on our system. The **base** and **boot** packages were automatically installed. We installed the **bootstrap** package [31] through the Packages menu. Although there is excellent support for bootstrap in the **boot** package, we also want to have access to the data sets for the book *An Introduction to the Bootstrap* by Efron and Tibshirani [14], which are available in the **bootstrap** package.

*Example 1.14 (Using the **bootstrap** package).* Suppose that we are interested in the examples related to the "law" data in the **bootstrap** package. Typing **law** at the prompt produces the following error because no object named "law" is found on the R search path.

```
> law
Error: object 'law' not found
```

We get a similar warning with

```
> data(law)
Warning message:
In data(law) : data set 'law' not found
```

In order to use **law** we first install the **bootstrap** package. This can be done in the R GUI using the Packages menu, or by typing the command

```
install.packages("bootstrap")
```

The system will prompt the user to select a server:

```
--- Please select a CRAN mirror for use in this session ---
```

and after the server is selected, the package will be installed. The only time that the package needs to be re-installed is when a new version of R is installed. A list of data files in the **bootstrap** package is displayed by

```
> data(package="bootstrap")
```

However, to use the objects in the package, one first needs to load it using the **library** function,

```
> library(bootstrap)
```

and then the objects in the package will be available. Loading the package with the **library** function typically needs to be done once in each new R session (each time the program is opened). Another useful feature of the **library** function is to get help for a package: either of the following display a summary of the package.

```
library(help=bootstrap)
help(package=bootstrap)
```

Finally, we want to use the **law** data, which is now loaded and accessible in the R workspace.

```
> library(bootstrap)
> law
   LSAT GPA
1   576 339
2   635 330
3   558 281
4   578 303
5   666 344
6   580 307
7   555 300
8   661 343
9   651 336
10  605 313
11  653 312
12  575 274
13  545 276
14  572 288
15  594 296
```

For example, we may want to compute the sample means or the correlation between LSAT scores and GPA.

```
> mean(law)
    LSAT        GPA
600.2667 309.4667
> cor(law$LSAT, law$GPA)
[1] 0.7763745
```

After installing R, and periodically thereafter, one should update packages using **update.packages** or the Packages menu.

## Searching for data or methods in packages

With thousands of contributed packages available for R, it is likely that whatever method one would like to apply has been implemented in an R package.

Many well known data sets can also be found in R packages. To search any installed packages, use `help.search` or `??`. Keep in mind that `help.search` and `??` do not search outside of the current R installation.

For example, suppose that we are trying to find a data set on heights of fathers and sons. We can try `??height` but none of the hits seem relevant. Next we go to the R homepage at `www.r-project.org` and click on *Search*. Using the R site search we get several relevant hits. The data set `galton` in package `UsingR` [51] is relevant ("Galton's height data for parents and children") and we also find a link to the data set `father.son` in `UsingR`, which is exactly what we were looking for. We can install the package `UsingR` and the `father.son` data will be available after loading the package with `library(UsingR)`.

One way to find an implementation of a particular method is to search on the R home page at `www.r-project.org`. A good general resource is to consult one or more of the *Task Views* (see page 38).

## 1.7 The R Workspace

When the user ends an R session by closing the R GUI or typing the quit command `q()`, a dialog appears asking "Save workspace image?". Usually one would not need to save the workspace image. Data is typically saved in files and reusable R commands should be saved in scripts.

One can list the objects in the current workspace by the command `ls()` or `objects()`. Starting with a new R session and no previously saved workspace, `ls()` returns an empty list.

```
> ls()
character(0)
```

After running the script "horsekicks.R" on page 11, a few objects have been added to the workspace. These objects will persist until the R session is ended or until removed or redefined by the user.

```
> source("/Rx/horsekicks.R")
[1] 0.545 0.325 0.110 0.015 0.005
[1] 0.61
[1] 0.6109548
     k   p           f
[1,] 0 0.545 0.543350869
[2,] 1 0.325 0.331444030
[3,] 2 0.110 0.101090429
[4,] 3 0.015 0.020555054
[5,] 4 0.005 0.003134646
```

The `ls` function displays the names of objects that now exist in the R workspace.

```
> ls()
[1] "f" "k" "p" "r" "v" "x"
```

To remove an object from the workspace use the rm or remove function.

```
> rm("v")
> ls()
[1] "f" "k" "p" "r" "x"

> remove(list=c("f", "r"))
> ls()
[1] "k" "p" "x"
```

If the workspace is saved upon exiting at this point, k, p, and x will be saved. However, saving a workspace combined with human error can lead to unnoticed serious programming errors. A typical example is when one forgets to define an object such as x but because it was already in the workspace (unintended and incorrect values for x) the R code may run without any reported errors. The unsuspecting user may never know that the analysis was completely wrong.

Finally to restore the R workspace to the "fresh" condition with no user-defined objects in the workspace, we can use the following code. This removes **all** objects listed by ls() without warning.

```
  rm(list = ls())
```

This is best done at the beginning or the end of a session. Wait until the end of this chapter to try it.

## 1.8 Options and Resources

### *The* options *function*

For more readable tables of data, we may want to round the displayed data. This can be done by explicitly rounding what is to be displayed:

```
> pi
[1] 3.141593
> round(pi, 5)
[1] 3.14159
```

Alternately, there is an option that governs the number of digits to display that can be set. The default is digits=7.

```
> options(digits=4)
> pi
[1] 3.142
```

To see the current values of options, type options(). Another option that helps to control the display is width; it controls how many characters are printed on each line. Illustrated below are two ways to get the current value of the width option, and changing the width option.

```
> options()$width        #current option for width
[1] 70
> options(width=60)      #change width to 60 characters
> getOption("width")     #current option for width
[1] 60
```

## *Graphical parameters:* par

Another set of options, for graphical parameters, is controlled by the **par** function. A useful one to know is how to change the "prompt user for next graph" behavior. It can be turned on/off by changing the graphics parameter **ask**; for example, to turn this prompt off:

```
par(ask = FALSE)
```

Another option that we use in this book is mfrow or mfcol to control how many figures are displayed in the current graphics window. For example, par(mfrow=c(2, 2)) will present figures in a 2 by 2 array, by row. The graphical parameters that can be set using **par** are described in the help topic ?par.

**R**$_\mathbf{x}$ **1.13** *There are so many possible parameters to graphics functions, that usually only a subset of them are listed in the documentation. For example, the* plot *help page starts with*

```
Generic function for plotting of R objects.  For more
details about the graphical parameter arguments, see 'par'.
```

*The* par *help page contains further documentation for* plot *and other graphics functions.*

## *Graph history*

In Windows, when a graphics window is active (on top), one can select *Recording* from the *History* menu. This has the effect of storing any graphs that are subsequently displayed, so that the user can use the page up/page down keys to page through the graphs.

If you construct multiple plots on a Macintosh, then with the graph selected, you can go Back or Forward from the Quartz menu to see previous graphs.

## *Other resources*

In addition to the manuals, frequently asked questions (FAQ), and online help included with R, there are many contributed files and web pages with excellent tutorials, examples, and explanations. A list is available on the R web site.[4]

### Task Views

"Task Views" for different types of statistical analyses are available on CRAN. Go to the R project home page at `www.r-project.org` and click on CRAN, then choose a mirror site near you. The CRAN page has a link to Task Views on several subjects. A Task View lists functions and packages that are related to the named task, such as "Bayesian", "Multivariate" or "Time Series". A direct link is `http://cran.at.r-project.org/web/views/`.

### External resources

In addition to materials found on the R project website, there are many useful materials to be found on the web. There is an interesting collection of information and examples in R Wiki,[5] including the list of examples in the R Graph Gallery. There is also a list of other R Wiki's. A nicely organized external resource is "Quick R: for SAS/SPSS/Stata Users" at `http://www.statmethods.net/index.html`.

### The R Graph Gallery and R Graphical Manual

For more experienced R users, a great resource for graphics is the *R Graph Gallery*.[6] We display the Gallery's home page and click on 'Thumbnails' to view small images of the graphs. Each graph includes the corresponding R code to produce the graph. Alternately one can select graphs by keyword or simply browse. The *R Graphical Manual*[7] illustrates thousands of graphs organized by image, task view, data set, or package.

---

[4] Contributed documentation, `http://cran.r-project.org/other-docs.html`

[5] R Wiki, `http://rwiki.sciviews.org/doku.php`

[6] R Graph Gallery, `http://addictedtor.free.fr/graphiques/`

[7] R Graphical Manual, `http://rgm2.lab.nig.ac.jp/RGM/images.php?show=all`

## 1.9 Reports and Reproducible Research

Most data analysis will be summarized in some type of report or article. The process of "copy and paste" for commands and output can lead to errors and omissions. Reproducible research refers to methods of reporting that combine the data analysis, output, graphics, and written report together in such a way that the entire analysis and report can be reproduced by others. Various formats for reports may include word processing documents, LATEX, or HTML.

The Sweave function in R facilitates generating this type of report. There is a LATEX package (*Sweave*) that generates a .tex file from the Sweave output. Various other packages such as *R2wd* (R to Word), *R2PPT* (R to PowerPoint), *odfWeave* (open document format), *R2HTML* (HTML), can be installed. Commercial packages are also available (e.g. *RTFGen* and *Inference for R*). For more details see the Task View "Reproducible Research" on CRAN.[8]

## Exercises

**1.1 (Normal percentiles).** The qnorm function returns the percentiles (quantiles) of a normal distribution. Use the qnorm function to find the $95^{th}$ percentile of the standard normal distribution. Then, use the qnorm function to find the quartiles of the standard normal distribution (the quartiles are the $25^{th}$, $50^{th}$, and $75^{th}$ percentiles). Hint: Use c(.25, .5, .75) as the first argument to qnorm.

**1.2 (Chi-square density curve).** Use the curve function to display the graph of the $\chi^2(1)$ density. The chi-square density function is dchisq.

**1.3 (Gamma densities).** Use the curve function to display the graph of the gamma density with shape parameter 1 and rate parameter 1. Then use the curve function with add=TRUE to display the graphs of the gamma density with shape parameter $k$ and rate 1 for 2,3, all in the same graphics window. The gamma density function is dgamma. Consult the help file ?dgamma to see how to specify the parameters.

**1.4 (Binomial probabilities).** Let $X$ be the number of "ones" obtained in 12 rolls of a fair die. Then $X$ has a Binomial($n = 12, p = 1/3$) distribution. Compute a table of binomial probabilities for $x = 0, 1, \ldots, 12$ by two methods:

a. Use the probability density formula

$$P(X = k) = \binom{n}{k} p^k (1-p)^{n-k}$$

---

[8] http://cran.at.r-project.org/web/views/ReproducibleResearch.html.

and vectorized arithmetic in R. Use `0:12` for the sequence of $x$ values and the `choose` function to compute the binomial coefficients $\binom{n}{k}$.

b. Use the `dbinom` function provided in R and compare your results using both methods.

**1.5 (Binomial CDF).** Let $X$ be the number of "ones" obtained in 12 rolls of a fair die. Then $X$ has a Binomial($n = 12, p = 1/3$) distribution. Compute a table of cumulative binomial probabilities (the CDF) for $x = 0, 1, \ldots, 12$ by two methods: (1) using `cumsum` and the result of Exercise 1.4, and (2) using the `pbinom` function. What is $P(X > 7)$?

**1.6 (Presidents' heights).** Refer to Example 1.2 where the heights of the United States Presidents are compared with their main opponent in the presidential election. Create a scatterplot of the loser's height vs the winner's height using the `plot` function. Compare the plot to the more detailed plot shown in the Wikipedia article "Heights of Presidents of the United States and presidential candidates" [54].

**1.7 (Simulated "horsekicks" data).** The `rpois` function generates random observations from a Poisson distribution. In Example 1.3, we compared the deaths due to horsekicks to a Poisson distribution with mean $\lambda = 0.61$, and in Example 1.4 we simulated random Poisson($\lambda = 0.61$) data. Use the `rpois` function to simulate very large ($n = 1000$ and $n = 10000$) Poisson($\lambda = 0.61$) random samples. Find the frequency distribution, mean and variance for the sample. Compare the theoretical Poisson density with the sample proportions (see Example 1.4).

**1.8 (*horsekicks*, continued).** Refer to Example 1.3. Using the `ppois` function, compute the cumulative distribution function (CDF) for the Poisson distribution with mean $\lambda = 0.61$, for the values 0 to 4. Compare these probabilities with the empirical CDF. The empirical CDF is the cumulative sum of the sample proportions `p`, which is easily computed using the `cumsum` function. Combine the values of `0:4`, the CDF, and the empirical CDF in a matrix to display these results in a single table.

**1.9 (Custom standard deviation function).** Write a function `sd.n` similar to the function `var.n` in Example 1.5 that will return the estimate $\hat{\sigma}$ (the square root of $\hat{\sigma}^2$). Try this function on the temperature data of Example 1.1.

**1.10 (Euclidean norm function).** Write a function `norm` that will compute the Euclidean norm of a numeric vector. The Euclidean norm of a vector $x = (x_1, \ldots, x_n)$ is

$$\|x\| = \sqrt{\sum_{i=1}^{n} x_i^2}.$$

Use vectorized operations to compute the sum. Try this function on the vectors $(0, 0, 0, 1)$ and $(2, 5, 2, 4)$ to check that your function result is correct.

**1.11 (Numerical integration).** Use the `curve` function to display the graph of the function $f(x) = e^{-x^2}/(1+x^2)$ on the interval $0 \le x \le 10$. Then use the `integrate` function to compute the value of the integral

$$\int_0^\infty \frac{e^{-x^2}}{1+x^2}\, dx.$$

The upper limit at infinity is specified by `upper=Inf` in the `integrate` function.

**1.12 (Bivariate normal).** Construct a matrix with 10 rows and 2 columns, containing random standard normal data:

```
x = matrix(rnorm(20), 10, 2)
```

This is a random sample of 10 observations from a standard bivariate normal distribution. Use the `apply` function and your `norm` function from Exercise 1.10 to compute the Euclidean norms for each of these 10 observations.

**1.13 (*lunatics* data).** Obtain a five-number summary for the numeric variables in the *lunatics* data set (see Example 1.12). From the summary we can get an idea about the skewness of variables by comparing the median and the mean population. Which of the distributions are skewed, and in which direction?

**1.14 (Tearing factor of paper).** The following data describe the tearing factor of paper manufactured under different pressures during pressing. The data is given in Hand et al. [21, Page 4]. Four sheets of paper were selected and tested from each of the five batches manufactured.

| Pressure | Tear factor | | | |
|---|---|---|---|---|
| 35.0 | 112 | 119 | 117 | 113 |
| 49.5 | 108 | 99 | 112 | 118 |
| 70.0 | 120 | 106 | 102 | 109 |
| 99.0 | 110 | 101 | 99 | 104 |
| 140.0 | 100 | 102 | 96 | 101 |

Enter this data into an R data frame with two variables: tear factor and pressure. Hint: it may be easiest to enter it into a spreadsheet, and then save it as a tab or comma delimited file (.txt or .csv). There should be 20 observations after a successful import.

**1.15 (Vectorized operations).** We have seen two examples of vectorized arithmetic in Example 1.1. In the conversion to Celsius, the operations involved one vector `temps` of length four and scalars (32 and 5/9). When we computed differences, the operation involved two vectors `temps` and `CT` of length four. In both cases, the arithmetic operations were applied element by element. What would happen if two vectors of different lengths appear

together in an arithmetic expression? Try the following examples using the
colon operator : to generate a sequence of consecutive integers.

a.

```
x = 1:8
n = 1:2
x + n
```

b.

```
n = 1:3
x + n
```

Explain how the elements of the shorter vector were "recycled" in each case.

# Chapter 2
# Quantitative Data

## 2.1 Introduction

This chapter covers some basic numerical and graphical summaries of data. Different numerical summaries and graphical displays would be appropriate for different types of data. A variable may be classified as one of the following types,

- Quantitative (numeric or integer)
- Ordinal (ordered, like integers)
- Qualitative (categorical, nominal, or factor)

and a data frame may contain several variables of possibly different types. There may be some structure to the data, such as in time series data, which has a time index. In this chapter we present examples of selected numerical and graphical summaries for various types of data. Chapter 3 covers summaries of categorical data in more detail. A natural continuation of Chapters 2 and 3 might be Chapter 5, "Exploratory Data Analysis."

## 2.2 Bivariate Data: Two Quantitative Variables

Our first example is a bivariate data set with two numeric variables, the body and brain size of mammals. We use it to illustrate some basic statistics, graphics, and operations on the data.

## 2.2.1 Exploring the data

### Body and brain size of mammals

There are many data sets included with the R distribution. A list of the
available data sets can be displayed with the data() command. MASS [50] is
one of the *recommended packages* that is bundled with the base R package,
so it should already be installed with R. To use the data sets or functions in
MASS one first loads MASS by the command

```
> library(MASS)   #load the package
> data()          #display available datasets
```

After the MASS package is loaded, the data sets in MASS will be included in
the list of available datasets generated by the data() command.

*Example 2.1 (mammals).* In the result of the data() command, under the
heading *"Data sets in package MASS:"* there is a data set named mammals.
The command

```
> ?mammals
```

displays information about the mammals data. This data contains brain size
and body size for 62 mammals. Typing the name of the data set causes the
data to be printed at the console. It is rather long, so here we just display
the first few observations using head.

```
> head(mammals)
                body brain
Arctic fox      3.385  44.5
Owl monkey      0.480  15.5
Mountain beaver 1.350   8.1
Cow           465.000 423.0
Grey wolf      36.330 119.5
Goat           27.660 115.0
```

This data consists of two numeric variables, body and brain.

$\mathbf{R_x}$ **2.1** *In the display above it is not obvious whether* mammals *is a matrix
or a data frame. One way to check whether we have a matrix or a data frame
is:*

```
> is.matrix(mammals)
[1] FALSE
> is.data.frame(mammals)
[1] TRUE
```

*One could convert* mammals *to a matrix by* as.matrix(mammals) *if a matrix
would be required in an analysis.*

## Some basic statistics and plots

The `summary` method computes a five number summary and mean for each numeric variable in the data frame.

```
> summary(mammals)
      body                 brain
 Min.    :   0.005   Min.    :   0.14
 1st Qu.:   0.600    1st Qu.:   4.25
 Median :   3.342    Median :  17.25
 Mean    : 198.790   Mean    : 283.13
 3rd Qu.:  48.203    3rd Qu.: 166.00
 Max.    :6654.000   Max.    :5712.00
```

If there were any missing data, these would appear in the summary as `NA`. The five number summaries are difficult to interpret because there are some extremely large observations (the means exceed not only the medians but also the third quartiles). This is clear if we view the five number summaries as side-by-side boxplots.

```
> boxplot(mammals)
```

The boxplots are shown in Figure 2.1(a). A scatterplot helps to visualize the relation between two quantitative variables. For a data frame with exactly two numeric variables like `mammals`, a scatterplot is obtained with the default arguments of `plot`.

```
> plot(mammals)
```

The scatterplot, shown in Figure 2.2(a), is not as informative as it could be because of the scale. A log transformation may produce a more informative plot. We display the scatterplot for the full `mammals` data set (on log-log scale) in Figure 2.2(b). (The `log` function computes the natural logarithm.)

```
> plot(log(mammals$body), log(mammals$brain),
+    xlab="log(body)", ylab="log(brain)")
```

In the second `plot` command we have also added descriptive labels for the axes.

**R**x **2.2** *We have seen that* mammals *consists of two numeric variables,* body *and* brain*, so the operation* log(mammals) *applies the natural logarithm function to the two numeric variables, returning a data frame of two numeric variables (log(body), log(brain)). This is a convenient shortcut, but of course it would not work if the data frame contained variables for which the logarithm is undefined.*

The summaries for the logarithms of body size and brain size are

```
> summary(log(mammals))
      body                brain
 Min.    :-5.2983   Min.    :-1.966
 1st Qu.:-0.5203    1st Qu.: 1.442
```

```
Median : 1.2066    Median : 2.848
Mean   : 1.3375    Mean   : 3.140
3rd Qu.: 3.8639    3rd Qu.: 5.111
Max.   : 8.8030    Max.   : 8.650
```

and the corresponding side-by-side boxplots of the transformed data shown
in Figure 2.1(b) are obtained by

```
boxplot(log(mammals), names=c("log(body)", "log(brain)"))
```

The default labels on the boxplots in Figure 2.1(b) would have been the
variable names ("body", "brain"), so we added more descriptive labels to the
plot with **names**.

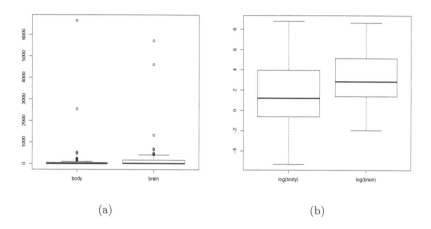

           (a)                                       (b)

**Fig. 2.1** Box plots of mammals brain size vs body size on (a) original and (b) log-log
scale in Example 2.1.

## 2.2.2 Correlation and regression line

In Figure 2.2(b) we can now observe a linear trend; logarithms of body and
brain size are positively correlated. We can compute the correlation matrix
of the data on the log-log scale by

```
> cor(log(mammals))
             body      brain
body   1.0000000 0.9595748
brain  0.9595748 1.0000000
```

or compute the correlation coefficient by

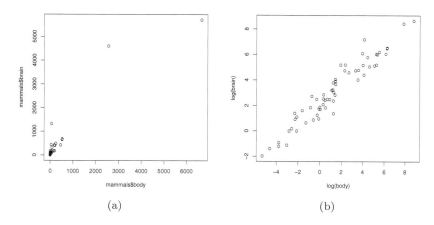

**Fig. 2.2** Scatterplots of mammals brain size vs body size on (a) original and (b) log-log scale in Example 2.1.

```
> cor(log(mammals$body), log(mammals$brain))
[1] 0.9595748
```

Simple linear regression is covered in Chapter 7. However, the code to add a fitted straight line to the log-log plot is quite simple. The `lm` function returns the coefficients of the line, and the line can be added to the log-log plot by the `abline` function (see Figure 2.3.)

```
> plot(log(mammals$body), log(mammals$brain),
+   xlab="log(body)", ylab="log(brain)")
> x = log(mammals$body); y = log(mammals$brain)
> abline(lm(y ~ x))
```

$\mathbf{R_x}$ **2.3** *A fitted line was added to the scatterplot in Figure 2.3 by the code* `abline(lm(y ~ x))`*. lm is a function and* `y ~ x` *is a* **formula***. This is one of many examples where the* **formula** *syntax is used in this book. A* **formula** *can be recognized by the tilde operator* $\sim$*, which connects a dependent variable on the left and predictor variable(s) on the right. Formulas will appear as arguments of some types of plot functions, and in the model specification argument for regression and analysis of variance model fitting. See the* **boxplot** *function that is used to produce Figure 2.4 below for an example of a formula argument in a plotting function.*

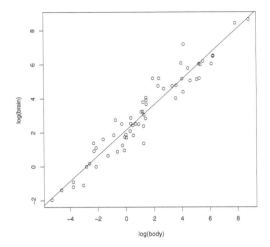

**Fig. 2.3** Scatterplot on log-log scale with fitted line in Example 2.1.

## 2.2.3 Analysis of bivariate data by group

**IQ of twins**

*Example 2.2 (IQ of twins separated near birth).* The data file "twinIQ.txt" contains IQ data on identical twins that were separated near birth. The data in the file is also available as a data set in the UsingR package [51] as twins. There are 27 observations on 3 variables:

| | |
|---|---|
| Foster | IQ for twin raised with foster parents |
| Biological | IQ for twin raised with biological parents |
| Social | Social status of biological parents |

The data set, which is shown in Table 2.1, can be imported using

```
> twins = read.table("c:/Rx/twinIQ.txt", header=TRUE)
```

and we display the first few observations with

```
> head(twins)
  Foster Biological Social
1     82         82   high
2     80         90   high
3     88         91   high
4    108        115   high
5    116        115   high
6    117        129   high
```

Next, we compute appropriate numerical summaries for each variable using `summary`.

```
> summary(twins)
      Foster          Biological         Social
 Min.   : 63.00   Min.   : 68.0   high  : 7
 1st Qu.: 84.50   1st Qu.: 83.5   low   :14
 Median : 94.00   Median : 94.0   middle: 6
 Mean   : 95.11   Mean   : 95.3
 3rd Qu.:107.50   3rd Qu.:104.5
 Max.   :132.00   Max.   :131.0
```

The `summary` function displays five number summaries and means for the two IQ scores, which are numeric, and a frequency table for the factor, `Social`. The five number summaries of the IQ scores are very similar.

**Table 2.1** IQ of twins separated near birth. The data is given in three columns in the file "twinIQ.txt".

| Foster | Biological | Social | Foster | Biological | Social | Foster | Biological | Social |
|--------|-----------|--------|--------|-----------|--------|--------|-----------|--------|
| 82 | 82 | high | 71 | 78 | middle | 63 | 68 | low |
| 80 | 90 | high | 75 | 79 | middle | 77 | 73 | low |
| 88 | 91 | high | 93 | 82 | middle | 86 | 81 | low |
| 108 | 115 | high | 95 | 97 | middle | 83 | 85 | low |
| 116 | 115 | high | 88 | 100 | middle | 93 | 87 | low |
| 117 | 129 | high | 111 | 107 | middle | 97 | 87 | low |
| 132 | 131 | high | | | | 87 | 93 | low |
| | | | | | | 94 | 94 | low |
| | | | | | | 96 | 95 | low |
| | | | | | | 112 | 97 | low |
| | | | | | | 113 | 97 | low |
| | | | | | | 106 | 103 | low |
| | | | | | | 107 | 106 | low |
| | | | | | | 98 | 111 | low |

We can display side-by-side boxplots of the difference in IQ scores by social status for comparison, using the formula `Foster - Biological ~ Social`.

```
> boxplot(Foster - Biological ~ Social, twins)
```

The boxplot shown in Figure 2.4 suggests that there could be differences in IQ for twins raised separately, but it is not clear whether the differences are significant. Another way to view this data is in a scatterplot, with the social status indicated by plotting character or color. This type of plot can be displayed by creating an integer variable for `Social` and using it to select the plotting characters (`pch`) and colors (`col`).

```
> status = as.integer(Social)
> status
 [1] 1 1 1 1 1 1 1 3 3 3 3 3 3 2 2 2 2 2 2 2 2 2 2 2 2 2 2
> plot(Foster ~ Biological, data=twins, pch=status, col=status)
```

The scatterplot is shown in Figure 2.5. Note that the levels of the factor `Social` are converted in alphabetical order: high=1, low=2, middle=3. To add the legend to the plot we used

```
> legend("topleft", c("high","low","middle"),
+    pch=1:3, col=1:3, inset=.02)
```

(On a color display, the symbols for *high*, *low*, and *medium* appear in colors black, red, and green, respectively.) To add the line Foster=Biological to the plot, we used `abline` with intercept 0 and slope 1:

```
> abline(0, 1)
```

Figure 2.5 does not reveal any dramatic differences by social status, although the high social status group may correspond to higher IQ scores for twins with their biological parents.

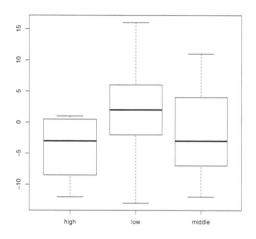

**Fig. 2.4** Boxplots of the differences in twins IQ scores (Foster-Biological) in Example 2.2.

## 2.2.4 Conditional plots

Continuing with the `twins` data, we illustrate a basic conditional plot displayed with the `coplot` function. Instead of displaying the data in different colors or plotting characters, `coplot` displays several scatterplots, all on the same scale. We set this up with a formula y $\sim$ x | a, which indicates that the plots of y vs x should be conditional on variable a.

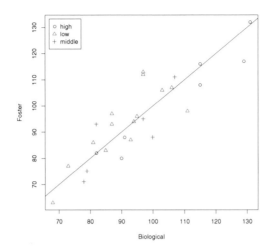

**Fig. 2.5** Scatterplot of twins IQ in Example 2.2.

```
> coplot(Foster ~ Biological|Social, data=twins)
```

The coplot is shown in Figure 2.6. The order of the plots is from the bottom and from the left (corresponding to increasing values of a, typically). In our coplot, Figure 2.6, that order is: high (lower left), low (lower right), middle (top left), because they are in alphabetical order.

Another version of this type of conditional plot is provided by the function xyplot in the lattice package. The lattice package is included in the R distribution so it should already be installed. To use the function xyplot we first load the lattice package using library. The basic syntax for xyplot in this example is

```
xyplot(Foster ~ Biological|Social, data=twins)
```

The above command displays a conditional plot (not shown) that is similar to the one in Figure 2.7, but with the default plotting character of an open blue circle. To obtain Figure 2.7 we used the following syntax that specifies a solid circle (pch=20) in the color black (col=1).

```
> library(lattice)
> xyplot(Foster ~ Biological|Social, data=twins, pch=20, col=1)
```

Neither of the conditional plots in Figures 2.6 or 2.7 reveal an obvious pattern or dependence on the conditioning variable social status.

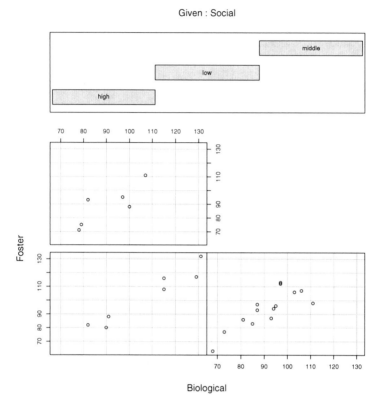

**Fig. 2.6** Conditional plot (`coplot`) of twins IQ by social status of biological parents, in Example 2.2.

## 2.3 Multivariate Data: Several Quantitative Variables

*Example 2.3 (Brain size and intelligence).*

Data from a study comparing brain size and intelligence is available on the DASL web site [12]. Willerman et al. [56] collected a sample of 40 students' IQ and brain size measured by MRI. There are 8 variables:

| Variable | Description |
| --- | --- |
| Gender | Male or Female |
| FSIQ | Full Scale IQ scores based on four Wechsler (1981) subtests |
| VIQ | Verbal IQ scores based on four Wechsler (1981) subtests |
| PIQ | Performance IQ scores based on four Wechsler (1981) subtests |
| Weight | Body weight in pounds |
| Height | Height in inches |
| MRI_Count | total pixel Count from the 18 MRI scans |

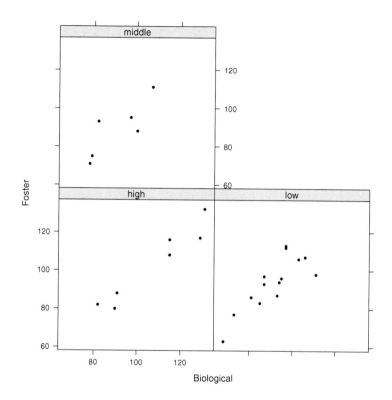

**Fig. 2.7** Conditional scatterplot (using `xyplot` in the `lattice` package) of twins IQ by social status of biological parents, in Example 2.2.

## 2.3.1 Exploring the data

After reading in the data using `read.table` we use the `summary` function to display summary statistics.

```
> brain = read.table("brainsize.txt", header=TRUE)
> summary(brain)
    Gender       FSIQ            VIQ             PIQ
 Female:20   Min.   : 77.00   Min.   : 71.0   Min.   : 72.00
 Male  :20   1st Qu.: 89.75   1st Qu.: 90.0   1st Qu.: 88.25
             Median :116.50   Median :113.0   Median :115.00
             Mean   :113.45   Mean   :112.3   Mean   :111.03
             3rd Qu.:135.50   3rd Qu.:129.8   3rd Qu.:128.00
             Max.   :144.00   Max.   :150.0   Max.   :150.00
```

```
      Weight              Height              MRI_Count
Min.    :106.0   Min.    :62.00   Min.    : 790619
1st Qu.:135.2    1st Qu.:66.00    1st Qu.: 855919
Median :146.5    Median :68.00    Median : 905399
Mean    :151.1   Mean    :68.53   Mean    : 908755
3rd Qu.:172.0    3rd Qu.:70.50    3rd Qu.: 950078
Max.    :192.0   Max.    :77.00   Max.    :1079549
NA's    :  2.0   NA's    : 1.00
```

The summary function displays appropriate summary statistics for each type of variable; a table is displayed for the categorical variable Gender, and the mean and quartiles are displayed for each numerical type of variable. The summaries show that there are some missing values of Weight and Height.

## 2.3.2 Missing values

There are several options for computing statistics for data with missing values. Many basic statistics functions such as mean, sd, or cor return missing values when missing values are present in the data. For example,

```
> mean(brain$Weight)
[1] NA
```

The help topic for a particular function explains what options are available. For the mean the optional argument is na.rm, which is FALSE by default. Setting it to TRUE allows the computation of the mean with the missing value(s) removed.

```
> mean(brain$Weight, na.rm=TRUE)
[1] 151.0526
```

This result is the same as the mean reported by summary on page 53.

## 2.3.3 Summarize by group

There are 20 males and 20 females in this data set, and on average males have larger bodies than females. Larger body size may be related to larger brain size, as we saw in Example 2.1. It may be informative to display statistics separately for males and females, which is easily done using the by function. The basic syntax for the by function is by(data, INDICES, FUN, ...) where the dots indicate possible additional arguments to the function named by FUN. This provides a place for our na.rm=TRUE argument to the mean function. For data we want to include all but the first variable using either brain[, -1] or brain[, 2:7]. This syntax, which omits the row index, indicates that we want all rows of data.

```
> by(data=brain[, -1], INDICES=brain$Gender, FUN=mean, na.rm=TRUE)
brain$Gender: Female
     FSIQ        VIQ        PIQ     Weight     Height  MRI_Count
  111.900    109.450    110.450    137.200     65.765 862654.600
-----------------------------------------------------------------
brain$Gender: Male
     FSIQ        VIQ        PIQ     Weight     Height   MRI_Count
 115.00000  115.25000  111.60000  166.44444   71.43158 954855.40000
```

As expected, the average weight, height, and MRI count is larger for males than for females.

A way to visualize the MRI counts by gender is to use a different color and/or plotting symbol in a scatterplot. The `plot` command is simpler if we first `attach` the data frame so that the variables can be referenced directly by name. A plot of `MRI_Count` by `Weight` is obtained by

```
> attach(brain)
> gender = as.integer(Gender)  #need integer for plot symbol, color
> plot(Weight, MRI_Count, pch=gender, col=gender)
```

It is helpful to add a legend to identify the symbol and color for each gender. The levels of `Gender` are numbered in alphabetical order when `gender` is created, so 1 indicates "Female" and 2 indicates "Male".

```
> legend("topleft", c("Female", "Male"), pch=1:2, col=1:2, inset=.02)
```

The plot with the legend is shown in Figure 2.8. In this plot it is easy to see an overall pattern that `MRI_Count` increases with `Weight` and that `MRI_Count` for males tend to be larger than `MRI_Count` for females.

## 2.3.4 Summarize pairs of variables

A pairs plot displays a scatterplot for each pair of quantitative variables. We want to display scatterplots for all pairs excluding the first variable (gender) in the data frame.

```
> pairs(brain[, 2:7])
```

In the pairs plot shown in Figure 2.9, each of the pairs of IQ plots have points that are clearly clustered in two groups. It appears that perhaps a group of lower IQ and higher IQ individuals were selected for this study. Consulting the online documentation[1] for this data, we find that: "With prior approval of the University's research review board, students selected for MRI were required to obtain prorated full-scale IQs of greater than 130 or less than 103, and were equally divided by sex and IQ classification."

---

[1] http://lib.stat.cmu.edu/DASL/Stories/BrainSizeandIntelligence.html

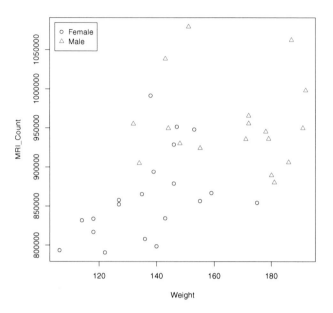

**Fig. 2.8** Scatterplot of MRI count vs weight in Example 2.3.

The pairs plot in Figure 2.9 reveals that some variables such as the IQ scores (FSIQ, VIQ, PIQ) have positive correlation. A table of Pearson correlation coefficients can be displayed using the `cor` function; here we round the table to two decimal places.

```
> round(cor(brain[, 2:7]), 2)
          FSIQ  VIQ  PIQ Weight Height MRI_Count
FSIQ      1.00 0.95 0.93     NA     NA      0.36
VIQ       0.95 1.00 0.78     NA     NA      0.34
PIQ       0.93 0.78 1.00     NA     NA      0.39
Weight      NA   NA   NA      1     NA        NA
Height      NA   NA   NA     NA      1        NA
MRI_Count 0.36 0.34 0.39     NA     NA      1.00
```

There are strong positive correlations between each of the IQ scores, but many of the correlations could not be computed due to the missing values in the data.

For computing covariances and correlations for data with missing values, there are several options. Here is one possible option that can be specified by the use argument in `cor`: if use="pairwise.complete.obs" then the correlation or covariance between each pair of variables is computed using all complete pairs of observations on those variables.

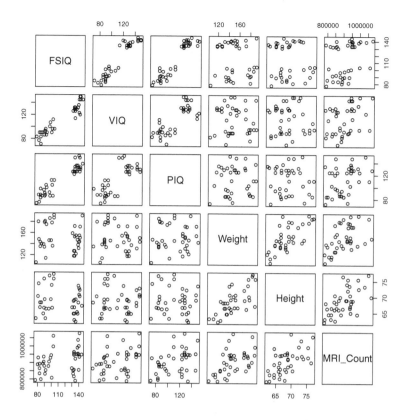

**Fig. 2.9** Pairs plot of brain size and IQ measurements in Example 2.3.

```
> round(cor(brain[, 2:7], use="pairwise.complete.obs"), 2)
          FSIQ   VIQ   PIQ Weight Height MRI_Count
FSIQ      1.00  0.95  0.93  -0.05  -0.09      0.36
VIQ       0.95  1.00  0.78  -0.08  -0.07      0.34
PIQ       0.93  0.78  1.00   0.00  -0.08      0.39
Weight   -0.05 -0.08  0.00   1.00   0.70      0.51
Height   -0.09 -0.07 -0.08   0.70   1.00      0.60
MRI_Count 0.36  0.34  0.39   0.51   0.60      1.00
```

(For another approach see the function `complete.cases`.)

The pairs plot (Figure 2.9) and the correlation matrix suggest a mild positive association ($r = 0.36$) between brain size and IQ. However, the MRI count is also correlated with body size (weight and height). If we control for body size as measured by say, weight,

```
> mri = MRI_Count / Weight
> cor(FSIQ, mri, use="pairwise.complete.obs")
[1] 0.2353080
```

the sample correlation of `mri` with `FSIQ` ($r = 0.235$) is smaller than the correlation of `MRI_Count` with `FSIQ` ($r = 0.36$).

One could test whether the correlation is significant using the correlation test `cor.test`. Before adjusting for body size, we have

```
> cor.test(FSIQ, MRI_Count)

        Pearson's product-moment correlation

data:  FSIQ and MRI_Count
t = 2.3608, df = 38, p-value = 0.02347
alternative hypothesis: true correlation is not equal to 0
95 percent confidence interval:
 0.05191544 0.60207414
sample estimates:
     cor
0.357641
```

The correlation test is significant at $\alpha = 0.05$. However, if we adjust for body size (using the transformed variable `mri`), the $p$-value is not significant at $\alpha = 0.05$.

```
> cor.test(FSIQ, mri)$p.value
[1] 0.1549858
```

For most statistical tests in R, the $p$-value of the test can be extracted like the example above.

## 2.3.5 Identifying missing values

This data frame (`brain`) has some missing values. The summary on page 53 indicates that there are two missing heights and one missing weight in the data. To identify these observations, we can use `which` and `is.na`.

```
> which(is.na(brain), arr.ind=TRUE)
     row col
[1,]   2   5
[2,]  21   5
[3,]  21   6
```

When using `which` on an object like a data frame or matrix we need `arr.ind=TRUE` to get the row and column numbers. Observations 2 and 21 have missing data in column 5 (height), and observation 21 has missing data in column 6 (weight). The missing observations are in rows 2 and 21, which we can extract by

```
> brain[c(2, 21), ]
   Gender FSIQ VIQ PIQ Weight Height MRI_Count
2    Male  140 150 124     NA   72.5   1001121
21   Male   83  83  86     NA     NA    892420
```

For example, one could replace missing values with the sample mean as follows.

```
> brain[2, 5] = mean(brain$Weight, na.rm=TRUE)
> brain[21, 5:6] = c(mean(brain$Weight, na.rm=TRUE),
+   mean(brain$Height, na.rm=TRUE))
```

The updated rows 2 and 21 in the data frame are

```
> brain[c(2, 22), ]
   Gender FSIQ VIQ PIQ  Weight Height MRI_Count
2    Male  140 150 124 151.0526   72.5   1001121
22   Male   97 107  84 186.0000   76.5    905940
```

## 2.4 Time Series Data

*Example 2.4 (New Haven temperatures).* The R data set `nhtemp` contains the average yearly temperatures in degrees Farenheit for New Haven, Connecticut, from 1912 to 1971. This is an example of a *time series*. The temperature variable is indexed by year.

```
> nhtemp
Time Series:
Start = 1912
End = 1971
Frequency = 1
 [1] 49.9 52.3 49.4 51.1 49.4 47.9 49.8 50.9 49.3 51.9 50.8 49.6
[13] 49.3 50.6 48.4 50.7 50.9 50.6 51.5 52.8 51.8 51.1 49.8 50.2
[25] 50.4 51.6 51.8 50.9 48.8 51.7 51.0 50.6 51.7 51.5 52.1 51.3
[37] 51.0 54.0 51.4 52.7 53.1 54.6 52.0 52.0 50.9 52.6 50.2 52.6
[49] 51.6 51.9 50.5 50.9 51.7 51.4 51.7 50.8 51.9 51.8 51.9 53.0
```

To visualize the pattern of temperatures over the years 1912 to 1971, we can easily display a time series plot using the `plot` function. For time series data, `plot` displays a line plot with time on the horizontal axis. The time plot is shown in Figure 2.10.

```
> plot(nhtemp)
```

One may be interested in identifying any trend in mean annual temperatures over the years represented by this data. One way to visualize possible trends is by fitting a smooth curve. One method of fitting a smooth curve is provided by the `lowess` function, which is based on locally-weighted polynomial regression. Below we plot the data again, this time including a more descriptive label for temperature, and add a horizontal reference line through the grand mean using `abline`. The smooth curve is added to the current plot with `lines`. (See Figure 2.11.)

```
> plot(nhtemp, ylab="Mean annual temperatures")
> abline(h = mean(nhtemp))
> lines(lowess(nhtemp))
```

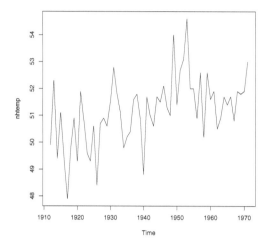

**Fig. 2.10** Time plot of mean annual temperatures in New Haven, Connecticut.

For modeling, one often wishes to transform a time series so that the mean is stable over time. When a mean appears to have an approximately linear trend over time as in Figure 2.11, first differences often remove the trend. If $X_1, X_2, \ldots$ is the time series, then we can obtain the time series of first differences $X_2 - X_1, X_3 - X_2, \ldots$ using the `diff` function.

```
> diff(nhtemp)
Time Series:
Start = 1913
End = 1971
Frequency = 1
 [1]  2.4 -2.9  1.7 -1.7 -1.5  1.9  1.1 -1.6  2.6 -1.1 -1.2 -0.3
[13]  1.3 -2.2  2.3  0.2 -0.3  0.9  1.3 -1.0 -0.7 -1.3  0.4  0.2
[25]  1.2  0.2 -0.9 -2.1  2.9 -0.7 -0.4  1.1 -0.2  0.6 -0.8 -0.3
[37]  3.0 -2.6  1.3  0.4  1.5 -2.6  0.0 -1.1  1.7 -2.4  2.4 -1.0
[49]  0.3 -1.4  0.4  0.8 -0.3  0.3 -0.9  1.1 -0.1  0.1  1.1
```

A time plot for the differenced series of temperatures with a reference line through 0 and `lowess` curve is obtained by the code below and shown in Figure 2.12.

```
> d = diff(nhtemp)
> plot(d, ylab="First differences of mean annual temperatures")
> abline(h = 0, lty=3)
> lines(lowess(d))
```

Figure 2.12 suggests that the mean of the differenced series is stable over time; notice that the lowess curve (solid line) is nearly horizontal and very close to the dotted horizontal line through 0.

## 2.5 Integer Data: Draft Lottery

*Example 2.5 (The 1970 Draft Lottery Data ).*

During the Vietnam War, the Selective Service System of the United States held a draft lottery on December 1, 1969 to determine the order of draft (induction) into the Army. The 366 birthdays (including leap year birthdays) were drawn to determine the order that eligible men would be called into service. Each birthday was matched with a number from 1 to 366 drawn from a barrel. The lowest numbers were drafted first. For an interesting discussion of statistical questions about this lottery see Fienberg [18] and Starr [46].[2]

The data and information about the lottery is available from the Selective Service System web site[3]. We converted it into a tab delimited file, which can be read into an R data frame by

```
> draftnums = read.table("draft-lottery.txt", header=TRUE)
```

This data frame is a table that contains the lottery numbers by day and month. The names of the variables are

```
> names(draftnums)
 [1] "Day" "Jan" "Feb" "Mar" "Apr" "May" "Jun" "Jul"
 [9] "Aug" "Sep" "Oct" "Nov" "Dec"
```

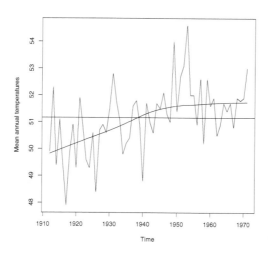

**Fig. 2.11** Time plot of mean annual temperatures in New Haven, Connecticut. A horizontal reference line is added to identify the grand mean and a smooth curve is added using `lowess`.

---

[2] http://www.amstat.org/publications/jse/v5n2/datasets.starr.html

[3] http://www.sss.gov/LOTTER8.HTM

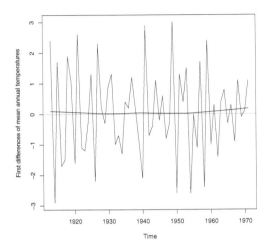

**Fig. 2.12** Time plot of first differences of mean annual temperatures in New Haven, Connecticut. A smooth curve is added using `lowess`.

To find the draft number for a January 15 birthday, for example, we read the 15th observation of column "Jan",

```
> draftnums$Jan[15]
[1] 17
```

and find that the corresponding draft number is 17.

Assuming the numbers were drawn randomly, we might expect that the medians of draft numbers for each month were near the number $366/2 = 183$. To display a table of medians for the draft numbers by month we "apply" the median function to the months (columns). The `sapply` function is a 'user-friendly' version of `apply`. However, by default the `median` function returns a missing value if there are missing values in the data.

```
> months = draftnums[2:13]
> sapply(months, median)
Jan Feb Mar Apr May Jun Jul Aug Sep Oct Nov Dec
211  NA 256  NA 226  NA 188 145  NA 201  NA 100
```

$\mathbf{R_x}$ 2.4 *The syntax* `draftnums[2:13]` *extracts the second through thirteenth variable from the data frame* `draftnums`. *It would be equivalent to use the syntax* `draftnums[, 2:13]` *or* `draftnums[, -1]`.

Our data has missing values for months with less than 31 days, so we use `na.rm=TRUE` in the `median` function. In `sapply`, the extra argument to `median` is simply listed after the name of the function.

```
> sapply(months, median, na.rm=TRUE)
  Jan   Feb   Mar   Apr   May   Jun   Jul   Aug   Sep
211.0 210.0 256.0 225.0 226.0 207.5 188.0 145.0 168.0
  Oct   Nov   Dec
201.0 131.5 100.0
```

The sample medians by month are less uniformly near 183 than one might expect by chance. A time plot of the medians by month can be obtained as follows.

```
> medians = sapply(months, median, na.rm=TRUE)
> plot(medians, type="b", xlab="month number")
```

In this plot we used `type="b"` to obtain both points and lines, and added a descriptive label "month number" on the horizontal axis. The plot (Figure 2.13) reveals an overall decreasing trend by month.

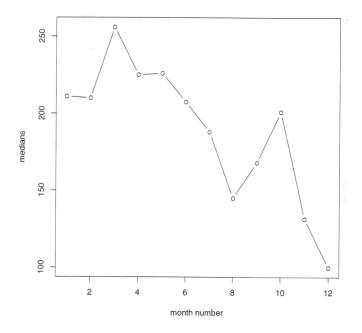

**Fig. 2.13** Medians of 1970 draft lottery numbers by month of birthday.

$\mathbf{R}_{\mathbf{x}}$ **2.5** *The result of* sapply *in Example 2.5 is a vector of medians in the order of the columns (months) January, . . . , December. Then Figure 2.13 is a plot of a vector of data. When the* plot *function is used with a single variable, a time plot of the data is displayed, with an index variable shown along the* *horizontal axis. The index values 1 through 12 in this case correspond to the month numbers.*

A boxplot  by months of the draft numbers is helpful for comparing the distributions of numbers for birthdays in different months.

```
> months = draftnums[2:13]
> boxplot(months)
```

The parallel boxplots shown in Figure 2.14 look less uniformly distributed across months than we might expect due to chance. The December birthdays appear to have lower draft numbers than birthdays in some other months. In fact, the numbers in the last two months of the year seem to be lower than other months overall. For a possible explanation, more draft lottery data, and further discussion of the lottery, see Starr [46] and "The Vietnam Lotteries" at `http://www.sss.gov/lotter1.htm`.

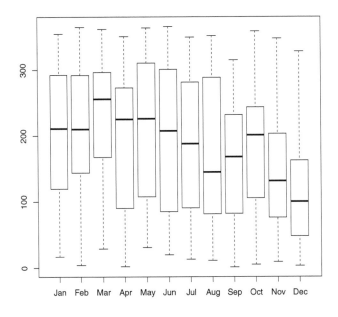

**Fig. 2.14** Boxplots comparing 1970 draft lottery numbers by month of birthday.

## 2.6 Sample Means and the Central Limit Theorem

*Example 2.6 (Sample means).* The data frame **randu** contains 400 triples of successive random numbers that were generated using an algorithm called RANDU. In this example we investigate the distribution of the sample mean of each triple of numbers. We know that if the numbers are truly from a continuous Uniform(0, 1) distribution, their expected value is $1/2$ and the variance is $1/12 = 0.08333$. First let us compute some basic sample statistics.

```
> mean(randu)
        x         y         z
0.5264293 0.4860531 0.4809547
> var(randu)
            x             y             z
x   0.081231885 -0.004057683  0.004637656
y  -0.004057683  0.086270206 -0.005148432
z   0.004637656 -0.005148432  0.077860433
```

Here, because **randu** is a data frame with three variables, the means are reported separately for each of the three variables, and the **var** function computes a variance-covariance matrix for the three variables. Each of the three sample means is close to the uniform mean $1/2$. The diagonal of the variance-covariance matrix has the three sample variances, which should be close to $0.08333$ under the assumption that RANDU generates Uniform(0, 1) data.

```
> diag(var(randu))
         x          y          z
0.08123189 0.08627021 0.07786043
```

The off-diagonal elements in the variance-covariance matrix are the sample covariance, and theoretically the covariances should be zero: the numbers in columns **x**, **y**, and **z** should be uncorrelated if in fact the RANDU table represents independent and identically distributed (iid) numbers. The sample correlations are each close to zero in absolute value:

```
> cor(randu)
            x           y           z
x   1.00000000 -0.04847127  0.05831454
y  -0.04847127  1.00000000 -0.06281830
z   0.05831454 -0.06281830  1.00000000
```

*Remark 2.1.* Although the **randu** data $(x, y, z)$ have correlations close to zero, in fact there is a linear relation that can be observed if we view the data in a 3-D plot. Try displaying the following plot to see that a pattern can be observed in the data (it is not quite random).

```
library(lattice)
cloud(z ~ x + y, data=randu)
```

See Chapter 4 for the plot (Figure 4.19, page 125) and further discussion of the `cloud` function.

We are interested in the distribution of the row means. Each row is assumed to be a random sample of size 3 from the Uniform(0,1) distribution. We can extract the row means using `apply`:

```
> means = apply(randu, MARGIN=1, FUN=mean)
```

Here `MARGIN=1` specifies rows and `FUN=mean` names the function to be applied to the rows. Alternately one could use

```
    rowMeans(randu)
```

to obtain the vector of means. Now `means` is a vector of 400 sample means. We plot a frequency histogram of the sample means using `hist`.

```
> hist(means)
```

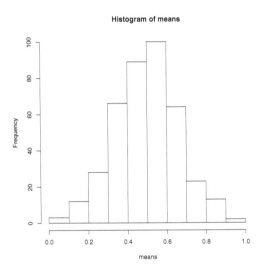

**Fig. 2.15** Frequency histogram produced with the default arguments to the `hist` function, for the sample means in Example 2.6.

The histogram of means shown in Figure 2.15 is mound shaped and somewhat symmetric. According to the Central Limit Theorem, the distribution of the sample mean tends to normal as the sample size tends to infinity. To compare our histogram with a normal distribution, however, we need a probability histogram, not a frequency histogram. A probability histogram (not shown) can be displayed by

```
> hist(means, prob=TRUE)
```

A density estimate can be displayed by:

```
> plot(density(means))
```

The density estimate is shown in Figure 2.16. It looks somewhat bell-shaped like a normal distribution.

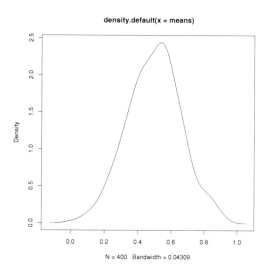

**Fig. 2.16** Density estimate for the sample means in Example 2.6.

By default the truehist function in the MASS package displays a probability histogram. We show the result of truehist below in Figure 2.17. Suppose we want to add a normal density to this histogram. The mean should be $1/2$ and the variance of the sample mean should be $\frac{1/12}{n} = \frac{1/12}{3} = 1/36$.

```
> truehist(means)
> curve(dnorm(x, 1/2, sd=sqrt(1/36)), add=TRUE)
```

From the histogram and normal density shown in Figure 2.17 one can observe that the distribution of sample means is approaching normality even with a sample size as small as three.

A normal-QQ plot compares the quantiles of a normal distribution with sample quantiles. When the sampled population is normal, the QQ plot should be close to a straight line. The plot and reference line are obtained by

```
> qqnorm(means)
> qqline(means)
```

The normal-QQ plot in Figure 2.18 is consistent with an approximately normal distribution of sample means.

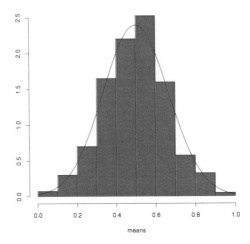

**Fig. 2.17** Histogram produced with `truehist` in `MASS` package for the sample means in Example 2.6. The normal density is added with `curve`.

## 2.7 Special Topics

### 2.7.1 Adding a new variable

*Example 2.7 (mammals, cont.).* In Example 2.1, suppose that we wish to create two categories of mammals, large and small. Say a mammal is "large" if body weight is above the median. A factor variable can be added to the data frame to indicate whether the mammal's body weight is above or below the median. We compute the median body size and use the `ifelse` function to assign "large" or "small" levels.

```
> m = median(mammals$body)
> mammals$size = ifelse(mammals$body >= m, "large", "small")
```

It is easy to understand `ifelse`; if the condition is true it returns the first value "large" and otherwise the second value "small".

$R_x$ **2.6** *The code*

```
mammals$size = ifelse(mammals$body >=m, "large", "small")
```

*assigns the "large" or "small" values to the variable* size *in the* mammals *data frame. Since the variable* size *does not yet exist in this data frame, a new variable* size *is created in this data frame.*

We use `head` to display the first six rows of the data frame:

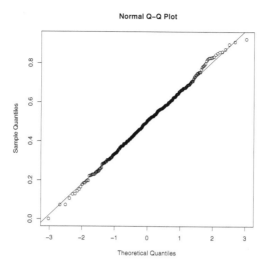

**Fig. 2.18** Normal-QQ plot of the sample means in Example 2.6.

```
> head(mammals)
                 body brain  size
Arctic fox      3.385  44.5 large
Owl monkey      0.480  15.5 small
Mountain beaver 1.350   8.1 small
Cow           465.000 423.0 large
Grey wolf      36.330 119.5 large
Goat           27.660 115.0 large
```

The new variable `size` makes it easy to carry out an analysis for the data separately for large and small mammals. For example,

```
subset(mammals, size=="large")
```

will return a data frame containing the large mammals only. The `==` operator is logical equality, not assignment.

## 2.7.2 Which observation is the maximum?

The variables `body` and `brain` can be referenced by `mammals$body` and `mammals$brain`. We can identify observation numbers using the `which` function. For example, here we identify the largest animals.

```
> which(mammals$body > 2000)
[1] 19 33

> mammals[c(19, 33), ]
```

```
                   body brain  size
Asian elephant     2547  4603 large
African elephant 6654  5712 large
```

The `which` function returned the row numbers of the mammals for which `mammals$body > 2000` is `TRUE`. Then we extracted these two rows from the data frame.

The maximum body size is

```
> max(mammals$body)
[1] 6654
```

Suppose that we want to identify the animal with the maximum body size, rather than the maximum value. A function that can be used to identify the largest mammal is `which.max`. It returns the index of the maximum, rather than the value of the maximum. The `which.min` function returns the index of the minimum. We can then use the results to extract the observations with the maximum or the minimum body size.

```
> which.max(mammals$body)
[1] 33
```

```
> mammals[33, ]
                  body brain  size
African elephant 6654  5712 large
```

```
> which.min(mammals$body)
[1] 14
```

```
> mammals[14, ]
                         body brain  size
Lesser short-tailed shrew 0.005  0.14 small
```

The African elephant has the greatest body size of 6654 kg, while the lesser short-tailed shrew has the smallest body size (0.005 kg) in this data set.

### 2.7.3 Sorting a data frame

*Example 2.8 (Sorting mammals).* Clearly the mammals are not listed in order of increasing body size. We could `sort` or `rank` the body size variable, but these functions do not help us to order the entire data frame according to body size. To list the mammals in a particular order, we first obtain the ordering using the `order` function. The `order` function will return a sequence of integers that sorts the data in the required order. To see how this works, let us take a small subset of the `mammals` data.

```
> x = mammals[1:5, ]    #the first five
> x
               body brain  size
Arctic fox     3.385 44.5 large
```

```
Owl monkey           0.480  15.5 small
Mountain beaver      1.350   8.1 small
Cow                465.000 423.0 large
Grey wolf           36.330 119.5 large
```

We want to sort the observations by body size.

```
> o = order(x$body)
> o
[1] 2 3 1 5 4
```

The result of `order` saved in the vector o indicates that in order of increasing body size we require observations 2, 3, 1, 5, 4. Finally using the result from `order` we re-index the data. We use the `[row, column]` syntax with o for the row and leave the column blank to indicate that we want all columns.

```
> x[o, ]
                 body brain  size
Owl monkey        0.480  15.5 small
Mountain beaver   1.350   8.1 small
Arctic fox        3.385  44.5 large
Grey wolf        36.330 119.5 large
Cow             465.000 423.0 large
```

The code to sort the full mammals data frame by body size is similar:

```
> o = order(mammals$body)
> sorted.data = mammals[o, ]
```

We display the last three observations of the sorted data frame using `tail`, and find the three largest body sizes with their corresponding brain sizes.

```
> tail(sorted.data, 3)
                  body brain  size
Giraffe            529   680 large
Asian elephant    2547  4603 large
African elephant  6654  5712 large
```

### 2.7.4 Distances between points

In this section we discuss the `dist` function for computing distances between points. We return to the `mammals` data introduced in Example 2.1 and continued in Examples 2.7-2.8, which contains body size and brain size of 62 mammals and a categorical variable `size` that we created in Example 2.7.

*Example 2.9 (Distances between points).* The original `mammals` data (body, brain) is an example of a bivariate (two-dimensional) data set. Distances between observations are defined only for numeric variables, so we begin by first reloading the data to restore the `mammals` data frame to its original form.

```
> data(mammals)
```

Suppose that we are interested in annotating the plot in Figure 2.2(b) with line segments representing the distances between some of the observations. The Euclidean distance between points $x = (x_1, \ldots, x_d)$ and $y = (y_1, \ldots, y_d)$ in a $d$-dimensional space is given by

$$\|x - y\| = \sqrt{\sum_{k=1}^{d} (x_k - y_k)^2}.$$

In two dimensions, as we have here, the distance is just the length of the line segment connecting the two points (the length of the hypotenuse of a right triangle). The distance matrix of a sample contains the distance between the $i^{th}$ and $j^{th}$ observation in row $i$, column $j$. A triangular table of distances is returned by `dist`; because the distance matrix is symmetric with zeroes on the diagonal, the `dist` object stores and displays it in a compact way. We illustrate with a small subset of `mammals`.

```
> x = mammals[1:5, ]

> dist(x)
                Arctic fox Owl monkey Mountain beaver        Cow
Owl monkey        29.145137
Mountain beaver   36.456841    7.450966
Cow              596.951136  617.928054      622.184324
Grey wolf         81.916867  110.005557      116.762838 525.233490
```

By default, the `dist` function returns Euclidean distances, but `dist` can optionally compute other types of distances as specified by the `method` argument.

For many applications, we require the full distance matrix. The `as.matrix` function converts a distance object to a matrix.

```
> as.matrix(dist(x))
                Arctic fox Owl monkey Mountain beaver       Cow Grey wolf
Arctic fox         0.00000   29.145137       36.456841 596.9511  81.91687
Owl monkey        29.14514    0.000000        7.450966 617.9281 110.00556
Mountain beaver   36.45684    7.450966        0.000000 622.1843 116.76284
Cow              596.95114  617.928054      622.184324   0.0000 525.23349
Grey wolf         81.91687  110.005557      116.762838 525.2335   0.00000
```

The scatterplot for the full `mammals` data frame (Figure 2.19) was created with this command:

```
> plot(log(mammals$body), log(mammals$brain),
+     xlab="log(body)", ylab="log(brain)")
```

Next, to display a few of the distances, we add line segments corresponding to the distances (cow, wolf), (wolf, human) and (cow, human). First we extract these three observations from `mammals` and store them in y. The three points form a triangle, so it is easy to draw the segments between them using `polygon`.

```
> y = log(mammals[c("Grey wolf", "Cow", "Human"), ])
> polygon(y)
```

(To add the segments one at a time, one could use the **segments** function.) Labels are added using **text**. The labels will be centered at coordinates of the points in y if we use:

```
    text(y, rownames(y))
```

The placement of the text labels can be adjusted. Here we used

```
> text(y, rownames(y), adj=c(1, .5))
```

for better placement of the labels. Also see Example 4.2 for an interactive method of labeling points using the **identify** function.

From the plot shown in Figure 2.19 we see that if measuring by the Euclidean distances of logarithms of brain and body size, humans are somewhat closer to cows than to wolves. The actual distances are

```
> dist(y)
          Grey wolf        Cow
Cow       2.845566
Human     2.460818 2.314068
```

## 2.7.5 Quick look at cluster analysis

*Example 2.10 (Cluster analysis of distances).* Distance matrices are often computed for cluster analysis. Cluster analysis is often applied to reveal possible structure in data. Although an in depth discussion of cluster analysis is beyond the scope of this book, in this section we take a quick look at how to implement a simple cluster analysis. A function that implements hierarchical cluster analysis is **hclust**. For example,

```
> d = dist(log(mammals))
> h = hclust(d, method="complete")
```

performs a hierarchical cluster analysis based on *furthest neighbors*. Using **method="complete"**, beginning with the singletons (individual observations), the two clusters with the smallest maximum pairwise distance are joined at each step. Clearly the distances in Figure 2.19 show that "cow" is not the nearest point to "cat" or to "human". Another widely applied method is *Ward's minimum variance*, obtained with the squared distance matrix and **method=ward** in hclust. Several other popular clustering methods are also implemented in **hclust**. A good reference on hierarchical cluster analysis is Everitt, Landau, and Leese [15].

In this example we will work with the largest half of the mammals. We extract the larger half by the **subset** function:

```
> big = subset(mammals, subset=(body > median(body)))
```

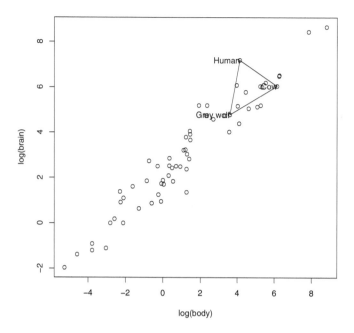

**Fig. 2.19** Scatterplot with distances between three of the mammals labeled in Example 2.1.

Then we compute distances and the clustering for the large animals:

```
> d = dist(log(big))
> h = hclust(d, method="complete")
```

The result of `hclust` can be plotted in a tree diagram called a *dendrogram*.

```
> plot(h)
```

The dendrogram is shown in Figure 2.20. The lowest branches to be clustered are the "nearest" according to this algorithm, while nodes that join at a higher level are less alike. For example, we see that the two elephants cluster together early, but are not joined with other clusters until much later. Note that this analysis is based entirely on brain and body sizes, so clusters represent relative size of the animal in some sense.

Let's see which pair of animals are the closest according to this clustering algorithm (the first pair to be merged into a cluster). That would be the pair with the smallest distance. That pair will be identified by the values returned in `h$merge`.

```
> head(h$merge)
      [,1] [,2]
[1,]  -22  -28
[2,]  -25  -57
[3,]  -21  -29
[4,]   -8  -12
[5,]   -9  -62
[6,]  -41  -60
```

We see that the logarithms of the 22nd and 28th observations have the smallest distance and therefore were the first cluster formed. These observations are:

```
> rownames(mammals)[c(22, 28)]
[1] "Horse"   "Giraffe"
```

and their logarithms are

```
> log(mammals)[c(22, 28), ]
            body     brain
Horse   6.255750 6.484635
Giraffe 6.270988 6.522093
```

Note that the cluster analysis will be different if the distances are computed on the original `mammals` data rather than the logarithms of the data, or if a different clustering algorithm is applied.

## Exercises

**2.1 (*chickwts* data).** The `chickwts` data are collected from an experiment to compare the effectiveness of various feed supplements on the growth rate of chickens (see `?chickwts`). The variables are `weight` gained by the chicks, and type of `feed`, a factor. Display side-by-side boxplots of the weights for the six different types of feeds, and interpret.

**2.2 (*iris* data).** The `iris` data gives the measurements in centimeters of the variables sepal length and width and petal length and width, respectively, for 50 flowers from each of three species of iris. There are four numeric variables corresponding to the sepal and petal measurements and one factor, `Species`. Display a table of means by `Species` (means should be computed separately for each of the three `Species`).

**2.3 (*mtcars* data).** Display the `mtcars` data included with R and read the documentation using `?mtcars`. Display parallel boxplots of the quantitative variables. Display a pairs plot of the quantitative variables. Does the pairs plot reveal any possible relations between the variables?

**2.4 (*mammals* data).** Refer to Example 2.7. Create a new variable $r$ equal to the ratio of brain size over body size. Using the full `mammals` data set, order

**Cluster Dendrogram**

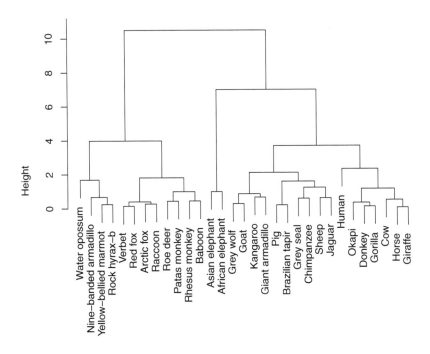

d
hclust (*, "complete")

**Fig. 2.20** Cluster dendrogram of log(mammals) data by nearest neighbor method in Example 2.1.

the `mammals` data by the ratio $r$. Which mammals have the largest ratios of brain size to body size? Which mammals have the smallest ratios? (Hint: use `head` and `tail` on the ordered data.)

**2.5 (*mammals* data, continued).** Refer to Exercise 2.5. Construct a scatterplot of the ratio $r = brain/body$ vs body size for the full `mammals` data set.

**2.6 (*LakeHuron* data).** The `LakeHuron` data contains annual measurements of the level, in feet, of Lake Huron from 1875 through 1972. Display a time plot of the data. Does the average level of the lake appear to be stable or changing with respect to time? Refer to Example 2.4 for a possible method of transforming this series so that the mean is stable, and plot the resulting series. Does the transformation help to stabilize the mean?

**2.7 (Central Limit Theorem with simulated data).** Refer to Example 2.6, where we computed sample means for each row of the `randu` data frame. Repeat the analysis, but instead of `randu`, create a matrix of random numbers using `runif`.

**2.8 (Central Limit Theorem, continued).** Refer to Example 2.6 and Exercise 2.7, where we computed sample means for each row of the data frame. Repeat the analysis in Exercise 2.7, but instead of sample size 3 generate a matrix that is 400 by 10 (sample size 10). Compare the histogram for sample size 3 and sample size 10. What does the Central Limit Theorem tell us about the distribution of the mean as sample size increases?

**2.9 (1970 Vietnam draft lottery).** What are some possible explanations for the apparent non-random patterns in the 1970 draft lottery numbers in Example 2.5? (See the references.)

**2.10 ("Old Faithful" histogram).** Use `hist` to display a *probability* histogram of the waiting times for the Old Faithful geyser in the *faithful* data set (see Example A.3). (Use the argument `prob=TRUE` or `freq=FALSE`.)

**2.11 ("Old Faithful" density estimate).** Use `hist` to display a *probability* histogram of the waiting times for the Old Faithful geyser in the *faithful* data set (see Example A.3) and add a `density` estimate using `lines`.

**2.12 (Ordering the *mammals* data by brain size).** Refer to Example 2.1. Using the full `mammals` data set, order the data by brain size. Which mammals have the largest brain sizes? Which mammals have the smallest brain sizes?

**2.13 (*mammals* data on original scale).** Refer to the `mammals` data in Example 2.7. Construct a scatterplot like Figure 2.19 on the original scale (Figure 2.19 is on the log-log scale.) Label the points and distances for cat, cow, and human. In this example, which plot is easier to interpret?

**2.14 (*mammals* cluster analysis).** Refer to Example 2.10. Repeat the cluster analysis using Ward's minimum variance method instead of nearest neighbor (complete) linkage. Ward's method is implemented in `hclust` with `method="ward"` when the first argument is the *squared* distance matrix. Display a dendrogram and compare the result with the dendrogram for the nearest neighbor method.

**2.15 (Identifying groups or clusters).** After cluster analysis, one is often interested in identifying groups or clusters in the data. In a hierarchical cluster analysis such as in Example 2.10, this corresponds to *cutting* the dendrogram (e.g. Figure 2.20) at a given level. The `cutree` function is an easy way to find the corresponding groups. For example, in Example 2.10, we saved the result of our complete-linkage clustering in an object `h`. To cut the tree to form five groups we use `cutree` with k=5:

```
g = cutree(h, 5)
```

Display g to see the labels of each observation. Summarize the group sizes using `table(g)`. There are three clusters that have only one mammal. Use `mammals[g > 2]` to identify which three mammals are singleton clusters.

# Chapter 3
# Categorical data

## 3.1 Introduction

In this chapter, we introduce R commands for organizing, summarizing, and displaying categorical data. We will see that categorical data is conveniently expressed by a special R object called a factor. The `table` function is useful in constructing frequency tables and the `plot` and `barplot` functions are useful in displaying tabulated output. The chi-square goodness-of-fit test for assessing if a vector of counts follows a specified discrete distribution is implemented in the `chisq.test` function. The `cut` function is helpful in dividing a numerical value into a categorical variable using a vector of dividing values. The `table` function with several variables can be used to construct a two-way frequency table and the `prop.table` function can be used to compute conditional proportions to explore the association pattern in the table. Side-by-side and segmented bar charts of conditional probabilities are constructed by the `barplot` function. The hypothesis of independence in a two-way table can be tested by the `chisq.test` function. A special graphical display `mosaicplot` can be used to display the counts in a two-way frequency table and, in addition, show the pattern of residuals from a fit of independence.

### 3.1.1 Tabulating and plotting categorical data

*Example 3.1 (Flipping a coin).*
    To begin, suppose we flip a coin 20 times and observe the sequence

$$H, T, H, H, T, H, H, T, H, H, T, T, H, T, T, T, H, H, H, T.$$

We are interested in tabulating these outcomes, finding the proportions of heads and tails, and graphing the proportions.

A convenient way of entering these data in the R console is with the `scan` function. One indicates by the `what=character` argument that character-type data will be entered. By default, this function assumes that "white space" will be separating the individual entries. We complete entering the outcomes by pressing the Enter key on a blank line. The character data is placed in the vector `tosses`.

```
> tosses = scan(what="character")
1: H T H H T H H T H H T T
13: H T T T
17: H H H T
21:
Read 20 items
```

We can tabulate this coin flipping data using the `table` function. The output is a table of frequencies of the different outcomes, H and T.

```
> table(tosses)
tosses
 H  T
11  9
```

We see that 11 heads and 9 tails were flipped. To summarize these counts, one typically computes proportions or relative frequencies. One can obtain these proportions by simply dividing the table frequencies by the number of flips using the `length` function.

```
> table(tosses) / length(tosses)
tosses
   H    T
0.55 0.45
```

There are several ways of displaying these data. First, we save the relative frequency output in the variable `prop.tosses`:

```
> prop.tosses = table(tosses) / length(tosses)
```

Using the `plot` method, we obtain a line graph displayed in Figure 3.1(a).

```
> plot(prop.tosses)
```

Alternately, one can display the proportions by a bar graph using the `barplot` function shown in Figure 3.1(b).

```
> barplot(prop.tosses)
```

The line graph and bar graph both give the same general impression that we obtained similar counts of heads and tails in this coin-tossing experiment.

### 3.1.2 Character vectors and factors

In the previous section, we illustrated the construction of a *character vector* which is a vector of string values such as "H" and "T." When one has a vector

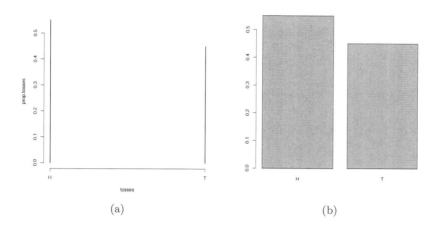

(a)                                                                (b)

**Fig. 3.1** Two displays of the proportions of heads and tails from 20 flips of a coin.

consisting of a small number of distinct values, either numerical or character, a *factor* is a useful way of representing this vector. To define a factor, we start with a vector of values, a second vector that gives the collection of possible values, and a third vector that gives labels to the possible values.

*Example 3.2 (Rolling a die).*

As a simple example, suppose we wish to collect several rolls of a die. We observe seven rolls of the die and save the rolls in the vector y.

```
> y = c(1, 4, 3, 5, 4, 2, 4)
```

For die rolls, we know that the possible rolls are 1 through 6 and we store these in the vector `possible.rolls`:

```
> possible.rolls = c(1, 2, 3, 4, 5, 6)
```

We wish to label the rolls by the words "one", ..., "six" – we place these labels in the vector `labels.rolls`:

```
> labels.rolls = c("one", "two", "three", "four", "five", "six")
```

We now are ready to construct the factor variable `fy` using the function `factor`:

```
> fy = factor(y, levels=possible.rolls, labels=labels.rolls)
```

By displaying the vector `fy`, we see the difference between a character vector and a factor.

```
> fy
[1] one    four  three five  four  two    four
Levels: one two three four five six
```

Note that the numerical roll values in y have been replaced by the factor labels in fy. Also, note that the display of the factor variable shows the *levels*, the possible values of the die roll. Suppose we construct a frequency table of the factor.

```
> table(fy)
fy
  one   two three  four  five   six
    1     1     1     3     1     0
```

Note that frequencies of all possible rolls of the die are displayed. In many situations, we wish to display the frequencies of categories such as "six" that are possible but have not been observed.

In the example to follow, a datafile is read that contains a character variable. When a data frame is created in R (say, using the read.table function), by default all variables consisting of character values are automatically converted to factors.

## 3.2 Chi-square Goodness-of-Fit Test

*Example 3.3 (Weldon's dice).*
"Weldon's Dice" is a famous data set published in Karl Pearson's 1900 paper [39] that introduced the chi-square goodness-of-fit test. At that time (before electronic computers) the English biologist Walter F. R. Weldon used dice to generate random data, recording the results of 26,306 rolls of 12 dice. See "Weldon's Dice, Automated" [29] for more details and results of an automated version of the experiment.

In the results, Weldon considered five or six dots showing among the 12 dice to be "successes" and other results "failures." If a single die is fair, then each face is equally likely to occur, so that the probability of a success (five or six) is 1/3. The total number of successes among 12 fair dice is a binomial random variable with success probability 1/3. The binomial probabilities are given by the dbinom function:

```
> k = 0:12
> p = dbinom(k, size=12, prob=1/3)
```

In 26,306 rolls of 12 fair dice, the expected outcomes (rounded to the nearest integer) would be:

```
> Binom = round(26306 * p)
> names(Binom) = k
```

Labels for the binomial counts were applied with the names function. The "Weldon's Dice" data are entered below.

```
> Weldon = c(185, 1149, 3265, 5475, 6114, 5194, 3067,
+    1331, 403, 105, 14, 4, 0)
> names(Weldon) = k
```

The binomial counts, data, and deviations between the binomial and observed counts can be summarized for display several ways; here we use the `data.frame` function. To combine data in a data frame we simply list the data vectors separated by commas. Here we also assigned a name "Diff" for the difference between the observed (Weldon) and expected (binomial) counts.

```
> data.frame(Binom, Weldon, Diff=Weldon - Binom)
   Binom Weldon Diff
0    203    185  -18
1   1216   1149  -67
2   3345   3265  -80
3   5576   5475 -101
4   6273   6114 -159
5   5018   5194  176
6   2927   3067  140
7   1255   1331   76
8    392    403   11
9     87    105   18
10    13     14    1
11     1      4    3
12     0      0    0
```

A visual comparison of the observed and expected counts can be made in several ways. To display the two bar plots of frequencies side by side we combine our data into a matrix using `cbind` and specify `beside=TRUE` in the `barplot` function.

```
> counts = cbind(Bin, Weldon)
> barplot(counts, beside=TRUE)
```

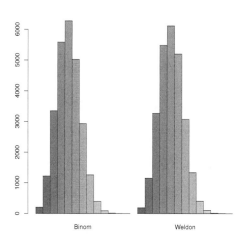

**Fig. 3.2** Bar plots of expected and observed frequencies in Weldon dice example.

The barplots in Figure 3.2 appear to agree approximately. For a comparison of observed and fitted data, however, it is somewhat easier to interpret the data on a single plot. Another possible plot for comparing the observed and expected counts is produced by:

```
> plot(k, Binom, type="h", lwd=2, lty=1, ylab="Count")
> lines(k + .2, Weldon, type="h", lwd=2, lty=2)
> legend(8, 5000, legend=c("Binomial", "Weldon"),
+      lty=c(1,2), lwd=c(2,2))
```

The extra arguments for plotting the binomial counts are `type="h"` (plot vertical lines like a barplot), `lwd=2` (line width is doubled), and `lty=1` (choose a solid line type). The `lines` function is used to overlay vertical lines for the corresponding observed counts. These lines are drawn slightly to the right of the expected count lines by adding a small value of 0.2 to the variable `k` and drawn using the alternative line type `lty=2`. This plot is shown in Figure 3.3.

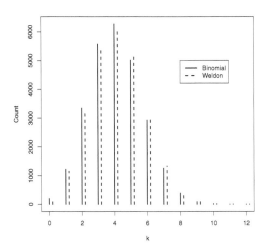

**Fig. 3.3** Line plot of expected and observed frequencies in Weldon dice example.

A chi-square goodness-of-fit test can be applied to test whether the data are consistent with the 'fair dice' binomial model. However, as we learn in basic statistics, the expected cell counts should be large — say, at least 5 for all cells. For a chi-square test we should collapse the categories corresponding to 10, 11, and 12 successes into a single category. This is accomplished with the following code.

```
>   cWeldon = c(Weldon[1:10], sum(Weldon[11:13]))
>   cWeldon
     0    1    2    3    4    5    6    7    8    9
   185 1149 3265 5475 6114 5194 3067 1331  403  105   18
```

One can now apply a chi-square goodness-of-fit test. We find the binomial probabilities for the first nine categories using the vector of probabilities stored in p. The sum of probabilities must equal 1, and this determines the probability for the final category. We use the `chisq.test` function to test the null hypothesis that the true model is binomial.

```
> probs = c(p[1:10], 1 - sum(p[1:10]))
> chisq.test(cWeldon, p=probs)
        Chi-squared test for given probabilities

data:  cWeldon
X-squared = 35.4943, df = 10, p-value = 0.0001028
```

The test computes a $p$-value of less than 0.001. Therefore, we conclude the experimental results that Weldon obtained are not consistent with the binomial model.

To better understand why the test rejects the binomial model, it is helpful to examine the Pearson residuals defined by

$$residual = \frac{observed - expected}{\sqrt{expected}},$$

where *observed* and *expected* denote respectively the observed and expected counts in a particular cell. To obtain these residuals, we save the results of the chi-square test in the variable `test` and the component `residuals` contains the vector of residuals. Using the `plot` function, we display the residuals as a function of k and overlay (using the `abline` function) a horizontal line at zero.

```
> test = chisq.test(cWeldon, p=probs)
> plot(0:10, test$residuals,
+   xlab="k", ylab="Residual")
> abline(h=0)
```

We see from Figure 3.4 that the observed Weldon counts are smaller than the binomial expected for values of k four or smaller, and larger than the expected counts for values of k larger than 4. One possible explanation for this conclusion is that the dice are not perfectly balanced.

## 3.3 Relating Two Categorical Variables

### 3.3.1 Introduction

*Example 3.4 (The twins dataset).*

Ashenfelter and Krueger [3] describe an interesting study to address the question "how much will an additional year of schooling raise one's income?" There are several difficulties in learning about the relationship be-

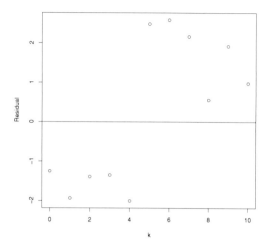

**Fig. 3.4** Graph of the Pearson residuals from the chi-square test for the Weldon example.

tween schooling and income. First, there are many variables besides schooling that relate to a person's income such as gender, socioeconomic status, and intelligence, and it is difficult to control for these other variables in this observational study. Second, it can be difficult to obtain truthful information about a person's schooling; people are more likely to report a higher level than they actually attain. The errors in obtaining actual educational levels can lead to biased estimates of the relationship between schooling and income. To address these concerns, these researchers interviewed twins, collecting information about income, education, and other background variables. Monozygotic twins (twins from one egg) have identical family backgrounds and they provide a good control for confounding variables. Also information about a person's education was obtained from a twin (self-reported) and also from his/her twin (cross-reported). By having two education measurements, one is able to estimate the bias from not getting a truthful response.

### 3.3.2 Frequency tables and graphs

We illustrate several statistical methods for categorical variable as a preliminary exploration of this twins dataset. We begin by reading in the dataset twins.dat.txt and storing it in the data frame twn:

```
> twn = read.table("twins.dat.txt", header=TRUE,
+     sep=",", na.strings=".")
```

There were 183 pairs of twins who were interviewed in this study. In each pair of twins, one twin was randomly assigned to "twin 1" and the other is called "twin 2." The variables EDUCL and EDUCH give the self-reported education (in years) for twin 1 and twin 2. (In the variable definition, the last letter of "L" refers to twin 1 and "H" refers to twin 2.) We can obtain frequency tables of the education years of the twins by two applications of the table function.

```
> table(twn$EDUCL)
 8 10 11 12 13 14 15 16 17 18 19 20
 1  4  1 61 21 30 11 37  1 10  3  3
> table(twn$EDUCH)
 8  9 10 11 12 13 14 15 16 17 18 19 20
 2  1  2  1 65 22 22 15 33  2 11  2  5
```

The education years for both twins show much variation and the year values with large frequencies are 12, corresponding to a high school degree, and 16, corresponding to a college degree.

Since there are so many educational year values, it is useful to categorize this variable into a smaller number of more meaningful levels. Suppose we say that a person's educational level is "high school" if he/she has 12 years of education, "some college" if the years are between 13 and 15, "college degree" for 16 years, and "graduate school" if the years are greater than 16. The cut function is very helpful for creating these new categories. In cut, the first argument is the variable to be changed, the argument breaks is a vector defining the breakpoints for the categories, and the argument labels is a character vector with the labels for the new categories. We use this function twice, once for twin 1's educational years and again for twin 2's educational years.

```
> c.EDUCL = cut(twn$EDUCL, breaks=c(0, 12, 15, 16, 24),
+   labels=c("High School", "Some College", "College Degree",
+   "Graduate School"))
> c.EDUCH = cut(twn$EDUCH, breaks=c(0, 12, 15, 16, 24),
+   labels=c("High School", "Some College", "College Degree",
+   "Graduate School"))
```

We tabulate the the educational levels for twin 1 using the table function, find relative frequencies by the prop.table function, and construct a bar graph of the relative frequencies using the barplot function. The resulting graph is shown in Figure 3.5. We see that approximately 70% of the first twins fall within the high school or some college levels.

```
> table(c.EDUCL)
c.EDUCL
    High School    Some College  College Degree Graduate School
             67              62              37              17
> prop.table(table(c.EDUCL))
c.EDUCL
    High School    Some College  College Degree Graduate School
     0.36612022      0.33879781      0.20218579      0.09289617
> barplot(prop.table(table(c.EDUCL)))
```

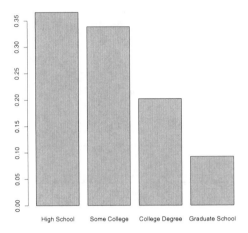

**Fig. 3.5** Bar graph of the educational levels for the first twin in the twins study

An alternative graphical display of table counts, a *mosaic plot*, is constructed using the function `mosaicplot`. In this display (shown in Figure 3.6), the total count, represented by a solid rectangle, is divided into vertical regions corresponding to the counts for the different educational levels. (Friendly [17] gives some history of this display.) Both the mosaic plot and the bar graph tell the same story – most of the first twins are in the high school and some college categories – but we will shortly see that the mosaic plot is especially helpful when we classify people with respect to two categorical variables.

```
> mosaicplot(table(c.EDUCL))
```

### 3.3.3 Contingency tables

The exploration of years of schooling in the previous section raises an interesting question. Is the educational level of twin 1 related to the educational level for twin 2? One can answer this question by constructing a contingency table of the educational levels for the two twins. This is easily constructed by the `table` function with the two variables `c.EDUCL` and `c.EDUCH` as arguments.

```
> table(c.EDUCL, c.EDUCH)
             c.EDUCH
c.EDUCL       High School Some College College Degree Graduate School
  High School          47           16              2               2
```

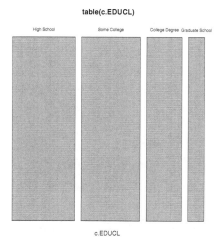

**Fig. 3.6** Mosaic plot of the educational levels for the first twin in the twins study

| | | | |
|---|---|---|---|
| Some College | 18 | 32 | 8 | 4 |
| College Degree | 5 | 10 | 18 | 4 |
| Graduate School | 1 | 1 | 5 | 10 |

Note from the contingency table that there are large counts along the diagonal of the table where the twins have the same self-reported educational levels. What proportion of twins have the same level? We store the table in the variable **T1** and extract the diagonal counts using the **diag** function:

```
> T1=table(c.EDUCL, c.EDUCH)
> diag(T1)
    High School    Some College  College Degree Graduate School
             47              32              18              10
```

We can compute the proportion of "same educational level" twins by two applications of the **sum** function, one on the vector of diagonal elements, and a second on the entire table.

```
> sum(diag(T1)) / sum(T1)
[1] 0.5846995
```

We see that about 58% of the twins have the same educational level.

One can graphically display the table by a mosaic plot constructed using the **plot** method for a table. (See Figure 3.7.)

```
> plot(T1)
```

This display is constructed in a two-step process. As in the first example, one first partitions a grey square into vertical bars where the widths of the bars correspond to the educational level counts for twin 1. Then each of the

vertical bars is divided into pieces where the heights of the pieces correspond
to the education counts for the second twin. The areas of the regions in the
mosaic plot correspond to the counts in the contingency table. Observe from
Figure 3.7 that the two largest areas correspond to the "high school, high
school" and "some college, some college" counts. This means that most of the
twins either both had a high school or both had "some college" background.

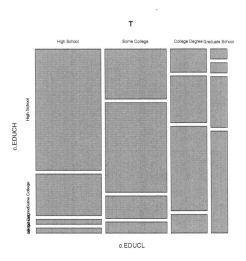

**Fig. 3.7** Mosaic plot of the educational levels for the twins in the twins study.

## 3.4 Association Patterns in Contingency Tables

### 3.4.1 Constructing a contingency table

The purpose of the twins study was to explore the relationship of educational
level with salary. We do some initial exploration by focusing on the data for
the first twin. The variable HRWAGEL contains the hourly wage (in dollars). If
one graphs the hourly wages by say, a histogram, one will observe the shape
of the wages is right-skewed. We divide the wages into four groups using the
cut function with break points 0, 7, 13, 20, and 150. We chose these break
points so we would have roughly the same number of twins in each class
interval. We assign the categorized wage to the variable c.wage.

```
> c.wage = cut(twn$HRWAGEL, c(0, 7, 13, 20, 150))
```

We construct a frequency table of the categorized wage by the table function.

```
> table(c.wage)
c.wage
  (0,7]    (7,13]   (13,20]  (20,150]
    47        58        38        19
```

There were 21 twins who did not respond to the wage question, so there were $183 - 21 = 162$ recorded wages.

To investigate the relationship of education with salary, we construct a two-way contingency table by another application of `table`; the first variable will appear as rows in the table and the second variable as columns.

```
> table(c.EDUCL, c.wage)
                c.wage
c.EDUCL          (0,7] (7,13] (13,20] (20,150]
  High School       23     21      10        1
  Some College      15     23      12        5
  College Degree     7     12      14        3
  Graduate School    2      2       2       10
```

We see that there were 23 people with a high school educational level and who are earning $7 or less per week, there were 21 people with a high school level and who are earning between $7 and $13 per week, and so on.

To quantify the relationship between education and salary, one can compute the proportion of different wage categories (column) for each educational level (row). This can be done using the `prop.table` function. The arguments are the table and the margin; if we use `margin = 1`, the proportions of each row of the table will be computed, and `margin = 2` the proportions of each column will be computed. Since we wish to compute proportions of different wages for each educational level, we first save the table in the variable T2, and use `prop.table` with arguments T2 and `margin = 1`.

```
> T2 = table(c.EDUCL, c.wage)
> prop.table(T2, margin=1)
                c.wage
c.EDUCL                  (0,7]      (7,13]     (13,20]    (20,150]
  High School      0.41818182 0.38181818 0.18181818 0.01818182
  Some College     0.27272727 0.41818182 0.21818182 0.09090909
  College Degree   0.19444444 0.33333333 0.38888889 0.08333333
  Graduate School  0.12500000 0.12500000 0.12500000 0.62500000
```

Of the high school students, we see from the table that 42% earned between 0 and $7, 32% earned between $7 and $13, and 18% earned between $13 and $20. Likewise, the table shows the proportions of students earning the different wage categories for each of the "Some College," "College Degree," and "Graduate School" educational levels.

## 3.4.2 Graphing patterns of association

There are several useful graphs for displaying the conditional proportions in
this contingency table. One way of displaying a proportion vector is a *segmented bar chart* where one divides a single bar into regions where the region
areas correspond to the proportions. The `barplot` function, when applied to
a matrix, will construct segmented bar charts for the column vectors of the
matrix. If `P` denotes the variable containing our proportion matrix, we are
interested in graphing the row (not column) vectors of `P`. So we first take the
transpose of `P` (using the `t` function) and then apply the `barplot` function.
We add several arguments in this function: we add a label of "PROPOR-
TION" along the y-axis, and add a legend that indicates the color of each of
the four wage categories. The resulting display is shown in Figure 3.8.

```
> P = prop.table(T2, 1)
> barplot(t(P), ylim=c(0, 1.3), ylab="PROPORTION",
+    legend.text=dimnames(P)$c.wage,
+    args.legend=list(x = "top"))
```

Note the areas of the lighter colored regions, corresponding to higher wages,
get larger from left to right, indicating that higher educational levels have
higher wages.

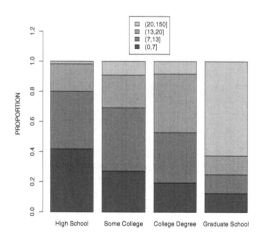

**Fig. 3.8** Segmented bar chart of the wage categories for people in four educational
levels.

Another useful display are side-by-side barplots, where the proportions
for a single educational level are displayed as a bar chart. This display is
constructed using the `barplot` using the `beside=TRUE` argument; see Figure

3.9. As in the previous graph, we add a legend that indicates the color of the bars corresponding to the four wage categories. Lower wages are colored using darker bars. We see that the darkest bars, corresponding to the lowest wage category, are predominant for the High School and Some College categories, and are unlikely for the Graduate College category.

```
> barplot(t(P), beside=T,  legend.text=dimnames(P)$c.wage,
+    args.legend=list(x="topleft"), ylab="PROPORTION")
```

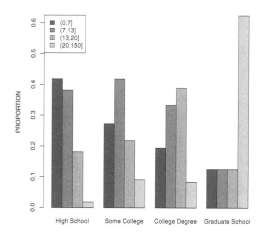

**Fig. 3.9** Side-by-side bar charts of the wage categories for people in four educational levels.

## 3.5 Testing Independence by a Chi-square Test

In the previous section, several methods for exploring the relationship between educational level and wage were illustrated. A more formal way of investigating the relationship between the two categorical variables is by a test of independence. If the variables educational level and wage are *independent*, this means that the probabilities of a twin earning the four wage categories will not depend on his/her education background. Based on our exploratory work, we strongly suspect that educational level and wage are *not* independent – a higher educational background appears to be associated with higher wages – but we will see that this statistical test will give new insight on how education and income are related.

The traditional test of independence is based on Pearson's chi-square statistic. If we test the hypothesis

$$H : \text{education background and wage are independent,}$$

we compute estimated expected counts under the assumption that $H$ is true. If we let *observed* denote the table of counts and *expected* denote the table of expected counts, then the Pearson statistic is defined by

$$X^2 = \sum_{\text{all cells}} \frac{(observed - expected)^2}{expected}.$$

The Pearson statistic measures the deviation of the observed counts from the expected counts and one rejects the hypothesis of independence for large values of the statistic $X^2$. If the hypothesis of independence is true, then $X^2$ will have, for large samples, an approximate chi-square distribution with degrees of freedom given by $df = (\text{number of rows} - 1) \times (\text{number of columns} - 1)$. Suppose that the computed value of $X^2$ for our data is equal to $X^2_{obs}$. The $p$-value is the probability of observing $X^2$ at least as extreme as $X^2_{obs}$; applying the chi-square approximation, this $p$-value is the probability that a chi-square$(df)$ random variable exceeds $X^2_{obs}$:

$$p - value = Prob(\chi^2_{df} \geq X^2_{obs}).$$

In R, recall we have the first twin's educational level stored in the variable c.EDUCL and the wage category stored in c.wage. The contingency table classifying twins by educational level and wage is stored in the variable T2:

```
> T2 = table(c.EDUCL, c.wage)
```

We perform a test of independence using the chisq.test function with the table T2 as the sole argument. We save the test calculations in the variable S.

```
> S  = chisq.test(T2)
Warning message:
In chisq.test(T2) : Chi-squared approximation may be incorrect
```

The warning tells that the accuracy of the chi-square approximation is in doubt due to a few small expected counts in the table. The results of the test are obtained by simply printing this variable.

```
> print(S)
        Pearson's Chi-squared test

data:  T2
X-squared = 54.5776, df = 9, p-value = 1.466e-08
```

It is instructive to confirm the calculations of this statistical test. One first computes the estimated expected counts of the table under the independence assumption – these expected counts are stored in the component `expected` of S that we display.

```
> S$expected
              c.wage
c.EDUCL                (0,7]     (7,13]    (13,20] (20,1e+03]
  High School      15.956790 19.691358 12.901235   6.450617
  Some College     15.956790 19.691358 12.901235   6.450617
  College Degree   10.444444 12.888889  8.444444   4.222222
  Graduate School   4.641975  5.728395  3.753086   1.876543
```

The observed counts are stored in the table T2. We can compute the test statistic by performing the operation "observed minus expected squared divided by expected" for all counts and summing over all cells.

```
> sum((T2 - S$expected)^2 / S$expected)
[1] 54.57759
```

Our answer, 54.57759, agrees with the displayed value of `X-squared` from the `chisq.test` output. Also we can check the computation of the $p$-value. The function `pchisq` computes the cdf of a chi-square random variable. Here the number of rows and number of columns of the table are both 4, and the degrees of freedom is equal to $df = (4-1)(4-1) = 9$. Since the distribution of the test statistic $X^2$ has approximately a chi-square(9) distribution under independence, the $p$-value is (approximately) the probability a $\chi^2(9)$ variate exceeds 54.57759 which is given by

```
> 1 - pchisq(54.57759, df=9)
[1] 1.465839e-08
```

This also agrees with the $p$-value given in the `chisq.test` output. This $p$-value is very small, so clearly the hypothesis of independence of educational level and wage category is rejected.

All of the calculations related to this chi-square test are stored in the variable S. One can view all components of S using the `names` function.

```
> names(S)
[1] "statistic" "parameter" "p.value"   "method"    "data.name" "observed"
[7] "expected"  "residuals"
```

One useful component is `residuals` – this contains the table of Pearson residuals, where a particular residual is defined by

$$residual = \frac{observed - expected}{\sqrt{expected}},$$

and *observed* and *expected* are, respectively, the count and the estimated expected count in that cell. By displaying the table of Pearson residuals, we see where the counts deviate from the independence model.

```
> S$residuals
              c.wage
c.EDUCL                (0,7]     (7,13]    (13,20]   (20,150]
```

```
High School       1.7631849  0.2949056 -0.8077318 -2.1460758
Some College     -0.2395212  0.7456104 -0.2509124 -0.5711527
College Degree   -1.0658020 -0.2475938  1.9117978 -0.5948119
Graduate School  -1.2262453 -1.5577776 -0.9049176  5.9300942
```

Informally, any residual larger than 2 in absolute value indicates a "significant" deviation from independence. It is interesting that using this criterion, there are two "large" residuals given in the rightmost column of the table. The residual of $-2.14$ indicates that there are fewer High School people earning wages over \$20 than anticipated by the independence model. In addition, the residual of 5.93 indicate there are more Graduate School people earning over \$20 that we would expect for independent variables. We can summarize the association by saying that educational level matters most in the highest wage category.

One can display the significant residuals by means of a mosaic plot. The mosaicplot function is first applied with the shade=FALSE (default) argument and the areas of the rectangles in the display in Figure 3.10(a) correspond to the counts in the table classifying twins by educational level and wage category.

```
> mosaicplot(T2, shade=FALSE)
```

If the shade=TRUE argument is used, one obtains an *extended* mosaic plot displayed in Figure 3.10(b). The border type and the shading of the rectangles relate to the sizes of the Pearson residuals. The two shaded rectangles correspond to the same two large residuals that we found by inspection of the table of residuals.

```
> mosaicplot(T2, shade=TRUE)
```

## Exercises

**3.1 (Fast food eating preference).** Fifteen students in a statistics class were asked to state their preference among the three restaurants Wendys, McDonalds, and Subway. The responses for the students are presented below.

```
Wendys    McDonalds Subway    Subway    Subway    Wendys
Wendys    Subway    Wendys    Subway    Subway    Subway
Subway    Subway    Subway
```

a. Use the scan function to read these data into the R command window.
b. Use the table function to find the frequencies of students who prefer the three restaurants.
c. Compute the proportions of students in each category.
d. Construct two different graphical displays of the proportions.

**3.2 (Dice rolls).** Suppose you roll a pair of dice 1000 times.

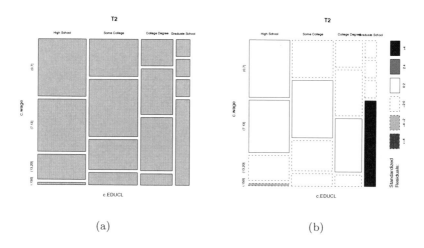

(a)                                              (b)

**Fig. 3.10** Mosaic plots of the table categorizing twin 1 by educational level and wage category. The left plot displays a basic mosaic plot and the right plot shows an extended mosaic plot where the shaded rectangles in the lower left and lower right sections of the graph correspond to large values of the corresponding Pearson residuals.

a. One can simulate 1000 rolls of a fair die using the R function `sample(6, 1000, replace=TRUE)`. Using this function twice, store 1000 simulated rolls of the first die in the variable `die1` and 1000 simulated rolls of the second die in the variable `die2`.

b. For each pair of rolls, compute the sum of rolls, and store the sums in the variable `die.sum`.

c. Use the `table` function to tabulate the values of the sum of die rolls. Compute the proportions for each sum value and compare these proportions with the exact probabilities of the sum of two die rolls.

**3.3 (Does baseball hitting data follow a binomial distribution?).** Albert Pujols is a baseball player who has $n$ opportunities to hit in a single game. If $y$ denotes the number of hits for a game, then it is reasonable to assume that $y$ has a binomial distribution with sample size $n$ and probability of success $p = 0.312$, where 0.312 is Pujols' batting average (success rate) for the 2010 baseball season.

a. In 70 games Pujols had exactly $n = 4$ opportunities to hit and the number of hits $y$ in these 70 games is tabulated in the following table. Use the `dbinom` function to compute the expected counts and the `chisq.test` function to test if the counts follow a binomial(4, 0.312) distribution.

b. In 25 games Pujols had exactly $n = 5$ opportunities to hit and the number of hits $y$ in these 25 games is shown in the table below. Use the `chisq.test` function to test if the counts follow a binomial(5, 0.312) distribution.

| Number of hits | 0 | 1 | 2 | 3 or more |
|---|---|---|---|---|
| Frequency | | 17 | 31 17 | 5 |

| Number of hits | 0 | 1 | 2 | 3 or more |
|---|---|---|---|---|
| Frequency | | 5 | 5 4 | 11 |

**3.4 (Categorizing ages in the twins dataset).** The variable AGE gives the age (in years) of twin 1.

a. Use the cut function on AGE with the breakpoints 30, 40, and 50 to create a categorized version of the twin's age.
b. Use the table function to find the frequencies in the four age categories.
c. Construct a graph of the proportions in the four age categories.

**3.5 (Relating age and wage in the twins dataset).** The variables AGE and HRWAGEL contain the age (in years) and hourly wage (in dollars) of twin 1.

a. Using two applications of the cut function, create a categorized version of AGE using the breakpoints 30, 40, and 50, and a categorized version of HRWAGEL using the same breakpoints as in Section 3.3.
b. Using the categorized versions of AGE and HRWAGEL, construct a contingency table of the two variables using the function table.
c. Use the prop.table function to find the proportions of twins in each age class that have the different wage groups.
d. Construct a suitable graph to show how the wage distribution depends on the age of the twin.
e. Use the conditional proportions in part (c) and the graph in part (d) to explain the relationship between age and wage of the twins.

**3.6 (Relating age and wage in the twins dataset, continued).**

a. Using the contingency table of the categorized version of AGE and HRWAGEL and the function chisq.test, perform a test of independence of age and wage. Based on this test, is there significant evidence to conclude that age and wage are dependent?
b. Compute and display the Pearson residuals from the test of independence. Find the residuals that exceed 2 in absolute value.
c. Use the function mosaicplot with the argument shade=TRUE to construct a mosaic plot of the table counts showing the extreme residuals.
d. Use the numerical and graphical work from parts (b) and (c) to explain how the table of age and wages differs from an independence structure.

**3.7 (Dice rolls, continued).** Suppose you roll a pair of dice 1000 times and you are interested in the relationship between the maximum of the two rolls and the sum of the rolls.

a. Using the `sample` function twice, simulate 1000 rolls of two dice and store the simulated rolls in the variables `die1` and `die2`.
b. The `pmax` function will return the parallel maximum value of two vectors. Using this function, compute the maximum for each of the 1000 pair of rolls and store the results in the vector `max.rolls`. Similarly, store the sum for each pair of rolls and store the sums in the vector `sum.rolls`.
c. Using the `table` function, construct a contingency table of the maximum roll and the sum of rolls.
d. By the computation of conditional proportions, explore the relationship between the maximum roll and the sum of rolls.

**3.8 (Are the digits of $\pi$ random?).** The National Institute of Standards and Technology has a web page that lists the first 5000 digits of the irrational number $\pi$. One can read these digits into R by means of the script

```
pidigits =
read.table("http://www.itl.nist.gov/div898/strd/univ/data/PiDigits.dat",
  skip=60)
```

a. Use the `table` function to construct a frequency table of the digits 1 through 9.
b. Construct a bar plot of the frequencies found in part (a).
c. Use the chi-square test, as implemented in the `chisq.test` function, to test the hypothesis that the digits 1 through 9 are equally probable in the digits of $\pi$.

# Chapter 4
# Presentation Graphics

## 4.1 Introduction

One of the most attractive aspects of the R system is its capability to produce state-of-the-art statistical graphics. All of the chapters of this book illustrate the application of a variety of R functions for graphing quantitative and categorical data for one or several dimensions. The purpose of this chapter is to describe methods for adjusting the attributes of the graph and for interacting with the graph that will enable the user to produce a publication-level graphical display. We focus on the methods that we have found useful in our own work.

The book by Murrell [37] provides a good description of the traditional graphics system. Sarkar [43] and Wickham [52] provide overviews respectively of the `lattice` and `ggplot2` packages for producing graphics that will be introduced at the end of this chapter.

*Example 4.1 (Home run hitting in baseball history).*

We begin with an interesting graph that helps us understand the history of professional baseball in the United States. Major League Baseball has been in existence since 1871 and counts of particular baseball events, such as the number of hits, doubles, and home runs, have been recorded for all of the years. One of the most exciting events in a baseball game is a home run (a batted ball typically hit out of the ballpark) and one may be interested in exploring the pattern of home run hitting over the years.

The data file "batting.history.txt" (extracted from the `baseball-reference.com` web site) contains various measures of hitting for each season of professional baseball. We read in the dataset and use the `attach` function to make the variables available for use.

```
> hitting.data = read.table("batting.history.txt", header=TRUE,
+   sep="\t")
> attach(hitting.data)
```

The variables `Year` and `HR` contain respectively the baseball season and number of home runs per game per team. We construct a time series plot of home runs against season using the `plot` function in Figure 4.1.

```
> plot(Year, HR)
```

Generally we see that home run hitting has increased over time, although there are more subtle patterns of change that we'll notice in the following sections.

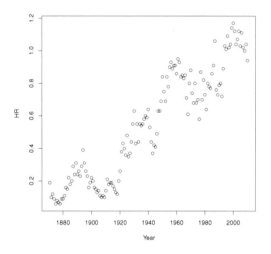

**Fig. 4.1** Display of the average number of home runs hit per game per team through all of the years of Major League Baseball.

## 4.2 Labeling the Axes and Adding a Title

To communicate a statistical graphic, it is important that the horizontal and vertical scales are given descriptive labels. In addition, it is helpful to give a graphic a good title so the reader understands the purpose of the graphical display. In this example, we think that "Season" is a better label than "Year" for the horizontal axis, and we want to more precisely define what HR means on the vertical axis. We add these labels using the `xlab` and `ylab` arguments to `plot` and we create a graph title using the `main` argument. Using these labels, we create a better display as shown in Figure 4.2.

```
> plot(Year, HR, xlab="Season", ylab="Avg HR Hit Per Team Per Game",
+    main="Home Run Hitting in the MLB Across Seasons")
```

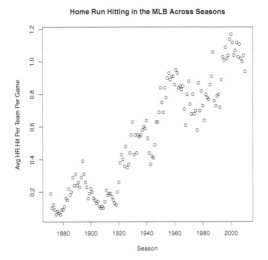

**Fig. 4.2** Home run hitting graph with axis labels and title added.

$\mathbf{R_x}$ **4.1** *By default, a high-level graphics function (such as* **plot***) will over-write the previous graph. One can avoid overwriting graphs by opening a new graphics window using the* **windows** *function on a Windows computer or* **quartz** *function on a Macintosh. Alternatively, after a graph is created, choose Recording from the History menu for a Windows interface. Then one can view different graphs by choosing "Next" and "Previous" from the History menu.*

## 4.3 Changing the Plot Type and Plotting Symbol

By default, the **plot** function produces a scatterplot of individually plotted points. Using the **type** argument option, one can connect the points with lines using **type = "l"**, show both points and connecting lines using **type = "b"**, or choose to show no points at all using **type = "n"**. In Figure 4.3, we illustrate the use of the **type = "b"** option which may be appropriate for some time series data.

```
> plot(Year, HR, xlab="Season", type="b",
+    ylab="Avg. Home Runs Hit by a Team in a Game",
+    main="Home Run Hitting in the MLB Across Seasons")
```

Using the **pch** argument in **plot**, one can specify the symbol of the plotting character. To see what symbols are available, Figure 4.4 draws all of the different plotting symbols together with the corresponding values of **pch**. The

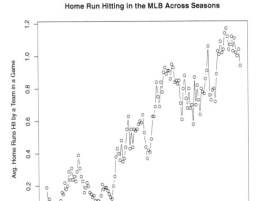

**Fig. 4.3** Home run hitting graph using `type` = `b` option.

following R code to produce Figure 4.4 illustrates how one can modify the usual plotting axes. In the `plot` function, the horizontal and vertical domains of the plotting region are controlled using the `xlim` and `ylim` arguments. The `type` = `"n"` option indicates that nothing is plotted, and the `xaxt` = `"n"` and `yaxt` = `"n"` arguments indicate that no axes are to be drawn. The labels on the x and y axes are suppressed using the `xlab` = `""` and `ylab` = `""` arguments. The different plotting symbols are overlaid on the current plot using the `points` function. The `row` and `col` vectors indicate the plotting position of the points, the `pch` argument indicate the plotting symbols to be drawn, and the `cex` = `3` argument magnifies the symbols to three times their usual size. Last, the `text` function allows one to display text on the current graph. Here we use the `text` function to overlay the numbers 0 through 20 next to the corresponding plotting symbols. These numbers are drawn 50% larger than their usual size (`cex` = `1.5`) and they are positioned to the right of the symbol (`pos` = `4`) about two character spaces away (`offset` = `2`). The `title` function allows one to add a title after the graph has been drawn.

```
> row = rep(1:3, each=7)
> col = rep(1:7, times=3)
> plot(2, 3, xlim=c(.5,3.5), ylim=c(.5,7.5),
+      type="n", xaxt = "n", yaxt = "n", xlab="", ylab="")
> points(row, col, pch=0:20, cex=3)
> text(row, col, 0:20, pos=4, offset=2, cex=1.5)
> title("Plotting Symbols with the pch Argument")
```

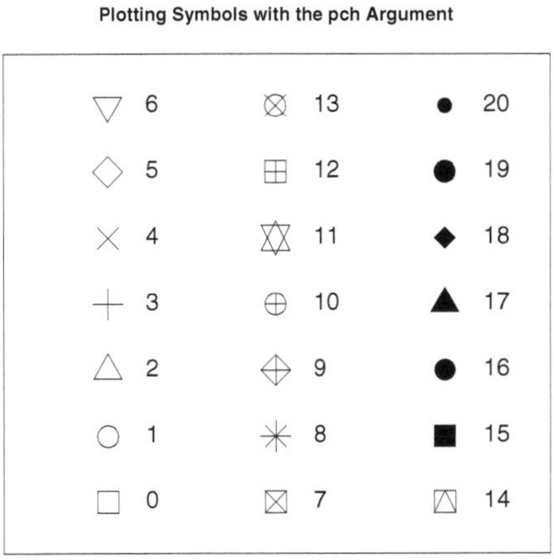

**Plotting Symbols with the pch Argument**

**Fig. 4.4** Display of 21 different shapes of plotting points using the `pch` argument option. The number next to the symbol shape is the corresponding value of `pch`.

Returning to our home run hitting graph, suppose we wish to use a solid circle plotting symbol (`pch = 19`) drawn 50% larger than default (`cex = 1.5`). Figure 4.5 shows the revised graph of the home run rates across seasons.

```
> plot(Year, HR, xlab="Season", cex=1.5, pch=19,
+    ylab="Avg. Home Runs Hit by a Team in a Game",
+    main="Home Run Hitting in the MLB Across Seasons")
```

## 4.4 Overlaying Lines and Line Types

To better understand the general pattern in a time series graph, it is helpful to apply a smoothing function to the data. A general, all-purpose *lowess* smoothing algorithm (Cleveland [10]) is implemented using the `lowess` function. In the following R code, the script `lowess(Year, HR)` will implement the lowess smoothing algorithm and the `lines` function overlays the smoothed points as a connected line on the current graph. One obtains the graph shown in Figure 4.6.

```
> plot(Year, HR, xlab="Season",
+    ylab="Avg. Home Runs Hit by a Team in a Game",
+    main="Home Run Hitting in the MLB Across Seasons")
> lines(lowess(Year, HR))
```

Looking at Figure 4.6, this smooth does not appear to be a good representation of the pattern of changes in the home run rates. The degree of

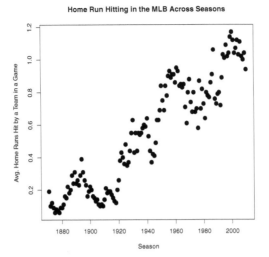

**Fig. 4.5** Home run hitting graph with the choices `pch = 19` and `cex = 1.5` arguments in the `plot` function.

smoothness in the lowess algorithm is controlled by the smoother span parameter $f$, and it seems that the default choice $f = 2/3$ has oversmoothed the data. This comment suggests that one should try several smaller values for the smoothness parameter $f$ and see the effect of these choices on the smooth on this particular dataset.

If one wishes to overlay several smoothing lines on the scatterplot, one should use different line patterns so that one can distinguish the different lines. One specifies the line style by the `lty` argument option. Six possible styles are numbered 1 through 6, that are equivalent to the strings "solid," "dashed," "dotted," "dotdash," "longdash," and "twodash." Figure 4.7 displays these six line styles using the following R code. As in the display of the point types, we first set up an empty graph with horizontal and vertical ranges from $-2$ to $2$ and suppress the printing of the axes and the axes labels. We draw lines with the `abline` function with slope `b = 1` and intercepts `a` of $2, 1, 0, -1, -2, -3$ with corresponding line styles 1 through 6. We use the `legend` function so the reader can match the line style with the values of `lty`. In the `legend` function, we indicate that the legend box is to be drawn in the "topleft" portion of the graph, the legend labels are to be the six values "solid, "dashed," "dotted," "dotdash," "longdash," and "twodash," and the `lty` and the `lwd` parameters indicate the line type and thickness of the lines associated with the labels. In this example, we decide to use thicker lines (`lwd = 2`), so that the five line types are distinguishable.

```
> plot(0, 0, type="n", xlim=c(-2, 2),  ylim=c(-2, 2),
+    xaxt="n", yaxt="n", xlab="", ylab="")
```

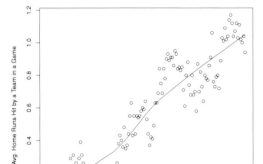

**Fig. 4.6** Home run hitting graph with overlaying lowess fit using the default smoothing parameter.

```
> y = seq(2, -3, -1)
> for(j in 1:6)
+   abline(a=y[j], b=1, lty=j, lwd=2)
> legend("topleft", legend=c("solid", "dashed", "dotted",
+   "dotdash", "longdash", "twodash"), lty=1:6, lwd=2)
> title("Line Styles with the lty Argument")
```

Using the different line styles, we will redraw our home run graph with three lowess fits corresponding to the smoothing parameter values $f = 2/3$ (the default value), $f = 1/3$, and $f = 1/12$ in Figure 4.8. After constructing the basic graph by `plot`, we use the `lines` function three times to overlay the smooths on the graph using the line style values `lty ="solid"` (the default), `lty = "dashed"`, and `lty = "dotdash"`. We add a legend, so the reader will be able to connect the line style with the smoothness parameter value. Comparing the three smooths, it appears that the lowess smooth with the value $f = 1/12$ is the best match to the pattern of increase and decrease in the scatterplot. From this smoothed curve, we see that home run hitting increased from 1940 to 1960, decreased from 1960 to 1980, and increased again from 1980 to 2000.

```
> plot(Year, HR, xlab="Season",
+   ylab="Avg. Home Runs Hit by a Team in a Game",
+   main="Home Run Hitting in the MLB Across Seasons")
> lines(lowess(Year, HR), lwd=2)
> lines(lowess(Year, HR, f=1 / 3), lty="dashed", lwd=2)
> lines(lowess(Year, HR, f=1 / 12), lty="dotdash", lwd=2)
```

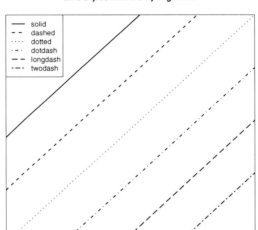

**Fig. 4.7** Display of six different line styles using the `lty` argument option.

```
> legend("topleft", legend=c("f = 2/3", "f = 1/3",
+     "f = 1/12"), lty=c(1, 2, 4), lwd=2,  inset=0.05)
```

## 4.5 Using Different Colors for Points and Lines

We have focused on the selection of different point shapes and line styles, but it may be preferable to distinguish points and lines using different colors. R has a large number of color choices that are accessible by the `col` argument to graphics functions such as `plot` or `hist`. One can appreciate the large number of available plotting colors by simply typing

```
> colors()
  [1] "white"            "aliceblue"         "antiquewhite"
  [4] "antiquewhite1"    "antiquewhite2"     "antiquewhite3"
  [7] "antiquewhite4"    "aquamarine"        "aquamarine1"
 [10] "aquamarine2"      "aquamarine3"       "aquamarine4"
 [13] "azure"            "azure1"            "azure2"
 [16] "azure3"           "azure4"            "beige"
 ...
```

As a simple example, suppose you wish to graph the following ten points $(1, 5), (2, 4), (3, 3), (4, 2), (5, 1), (6, 2), (7, 3), (8, 4), (9, 3), (10, 2)$ using

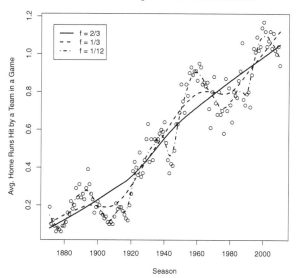

**Fig. 4.8** Home run hitting graph with three lowess fits with three choices of the smoothing parameter.

a variety of colors. Try the following R code (output not shown) to see the display of colors. To make the colors more visible, the `plot` function uses the (`cx = 5, pch = 19`) arguments to draw large solid points.

```
> plot(1:10, c(5, 4, 3, 2, 1, 2, 3, 4, 3, 2),
+   pch=19, cex=5,
+   col=c("red", "blue", "green", "beige", "goldenrod",
+       "turquoise", "salmon", "purple", "pink", "seashell"))
```

One can also specify colors by numbers. By typing the following R code (output not shown), one can see the default mapping of the numbers 1 through 8 to colors.

```
> palette()
[1] "black"    "red"      "green3"   "blue"     "cyan"     "magenta" "yellow"
[8] "gray"
```

The argument `col = 1` corresponds to black, `col = 2` corresponds to red, and so on. The following R code displays four lines drawn using the line styles `lty = 1`, ..., `lty = 4`. (The graph is not shown.)

```
> plot(0, 0, type="n", xlim=c(-2, 2), ylim=c(-2, 2),
+   xaxt="n", yaxt="n", xlab="", ylab="")
> y = c(-1, 1, 0, 50000)
```

```
> for (j in 1:4)
+   abline(a=0, b=y[j], lty=j, lwd=4)
```

If one adds the additional argument `col=j` to `abline`, one will see four lines drawn using contrasting line types and contrasting colors.

## 4.6 Changing the Format of Text

One has fine control over the style and placement of textual material placed on a graph. Specifically, one can choose the font family, style, and size of text through arguments of the `text` function. We illustrate some of these options through a simple example.

We begin by setting up an empty plot window where the horizontal limits are −1 and 6 and the vertical limits are −0.5 and 4. We place the text `"font = 1 (Default)"` at the coordinate point $(2.5, 4)$. As the label suggests, this text is drawn using the default font family and size. (See the top line of text in the box in Figure 4.9.)

```
> plot(0, 0, type="n", xlim=c(-1, 6), ylim=c(-0.5, 4),
+    xaxt="n", yaxt ="n", xlab="", ylab="",
+    main="Font Choices Using, font, family and srt Arguments")
> text(2.5, 4, "font = 1 (Default)")
```

One can change the style of the text font using the `font` argument option. The values `font = 2`, `font = 3`, and `font = 4` correspond respectively to bold, italic, and bold-italic styles. Using the `srt` argument, one can rotate the text string by an angle (in degrees) in a counter-clockwise direction. We illustrate rotating the last string 20 degrees. (See the three lines of text on the left side of the box in Figure 4.9.)

```
> text(1, 3, "font = 2 (Bold)", font=2, cex=1.0)
> text(1, 2, "font = 3 (Italic)", font=3, cex=1.0)
> text(1, 1, "font = 4 (Bold Italic), srt = 20", font=4,
+   cex=1.0, srt=20)
```

One can control the size of the text using the `cex` option. In the R code, we use the `cex = 1.0` which corresponds to the default size. One can double the text size using `cex = 2` or decrease the text size using `cex = 0.5`.

The argument `family` is used to select the font family of the text. One can choose the alternative serif, sans, and mono families by the use of the `"serif"`, `"sans"`, and `"mono"` values for `family`. The following R code illustrates using `text` with these three common families and Figure 4.9 shows the results on the right hand side inside the box. The Hershey font families are also available. In the last line of code, we illustrate using the `"HersheyScript"` family, drawn at two and a half times the usual size using cex = 2.5, and drawn in red using `col = "red"`.

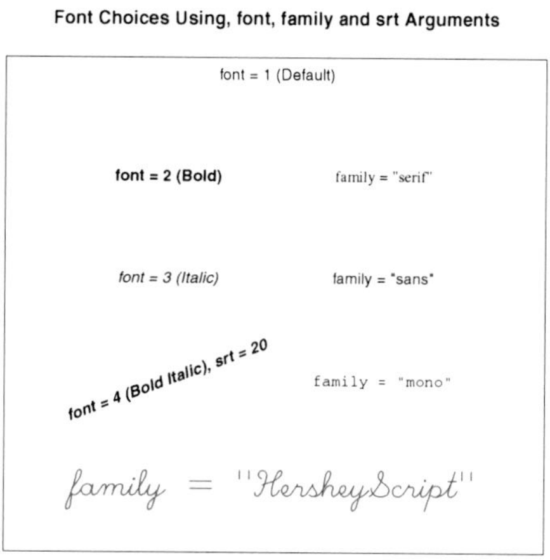

Font Choices Using, font, family and srt Arguments

font = 1 (Default)

**font = 2 (Bold)**          family = "serif"

*font = 3 (Italic)*          family = "sans"

*font = 4 (Bold Italic), srt = 20*          family = "mono"

*family = "HersheyScript"*

**Fig. 4.9** Illustration of the use of different font, family, and srt arguments in the text function.

```
> text(4, 3, 'family="serif"', cex=1.0, family="serif")
> text(4, 2, 'family="sans"', cex=1.0, family="sans")
> text(4, 1, 'family="mono"', cex=1.0, family="mono")
> text(2.5, 0, 'family = "HersheyScript"', cex=2.5,
+    family="HersheyScript", col="red")
```

## 4.7 Interacting with the Graph

One helpful function is identify that allows a person to interact with a graph. Here we use identify to label interesting points in a scatterplot.

*Example 4.2 (Home run hitting in baseball history (continued)).*
     We described the basic pattern in the home run hitting graph by a lowess smooth. We are next interested in examining the residuals, the vertical deviations of the points from the fitted points. In the following R script, we first save the lowess fit calculations in the variable fit. The component fit$y contains the fitted points and the residuals (actual HR values minus the fitted values) are computed and stored in the vector Residual. We construct a scatterplot of Residual against Year by the plot function and overlay a horizontal line at zero using the abline function.

```
> fit = lowess(Year, HR, f=1 / 12)
> Residual = HR - fit$y
> plot(Year, Residual)
> abline(h=0)
```

Looking at this graph (Figure 4.10), we notice two unusually large residuals, one positive and one negative. The `identify` function can be used to learn which seasons corresponded to these unusual residuals. This function has four arguments: the x and y plotting variables (`Year` and `Residual`), the number of points to identify (`n = 2`), and the vector of plotting labels to use (`Year`). After this function is executed, a crosshair appears when one moves the mouse over the graphics window. One identifies the points by clicking the mouse near the two points and the years corresponding to these points will appear. We see that the unusually large positive and large negative residuals correspond respectively to the 1987 and 1968 seasons. (The 1987 baseball season is notable in that players used steroids to help their hitting and the 1968 was a season where pitching was unusually dominant.) The output of `identify` is a vector with values 98 and 117 which are the row numbers of the data frame corresponding to these identified points.

```
> identify(Year, Residual, n=2, labels=Year)
[1]   98 117
```

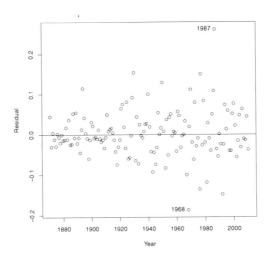

**Fig. 4.10** Residual graph from lowess fit with two outliers identified.

## 4.8 Multiple Figures in a Window

Sometimes it is desirable to put several graphs in the same plot window. In the traditional graphics system, this can be accomplished using the `mfrow` and `mfcol` parameter settings (the mnemonic is mf for "multiple figure"). Suppose one wishes to divide the window into six plotting regions in a grid with three rows and two columns. This will be set up by typing

```
> par(mfrow=c(3, 2))
```

Then a succession of six graphs, say using the `plot` function, will be displayed in the six plotting regions, where the the top regions are used first. If one wishes to restore the display back to a single plotting region, then type

```
> par(mfrow=c(1, 1))
```

In our example, suppose we wish to have two regions, where the home run rates and the lowess fit are displayed in the top region, and the residual graph is displayed in the bottom region. We begin using the `par` function with the `mfrow=c(2,1)` argument, and then follow with two `plot` functions displayed in the two regions (see Figure 4.11).

```
> par(mfrow=c(2, 1))
> plot(Year, HR, xlab="Season",
+    ylab="Avg HR Hit Per Team Per Game",
+    main="Home Run Hitting in the MLB Across Seasons")
> lines(fit, lwd=2)
> plot(Year, Residual, xlab="Season",
+    main="Residuals from Lowess Fit")
> abline(h=0)
```

## 4.9 Overlaying a Curve and Adding a Mathematical Expression

*Example 4.3 (Normal approximation to binomial probabilities).*
   We use a new example to illustrate a few additional useful R graphics functions. It is well known that binomial probabilities can be closely approximated by the normal distribution. We wish to demonstrate this fact graphically by displaying a set of binomial probabilities and overlaying the approximating normal curve.

   As our example, consider binomial probabilities for an experiment with sample size $n = 20$ and probability of success $p = 0.2$. We set up a grid of possible values for the binomial variable $y$, compute the corresponding probabilities using the R `dbinom` function, and store the probabilities in the vector `py`.

```
> n = 20; p = 0.2
> y = 0:20
> py = dbinom(y, size=n, prob=p)
```

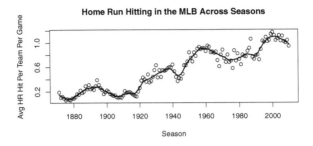

**Fig. 4.11** Home run hitting graph with fit and residual plot.

We construct a *histogram* or vertical line style display using `plot` with the `type = "h"` argument option. We decide to plot thick lines with the `lwd=3` option, and limit the horizontal plotting range to (0, 15) by the `xlim` argument. Using `ylab`, we give a descriptive label for the vertical axis. The resulting display is shown in Figure 4.12.

```
> plot(y, py, type="h", lwd=3,
+   xlim=c(0, 15), ylab="Prob(y)")
```

Next we wish to overlay a matching normal curve to this plot. We compute the mean and standard deviation of the binomial distribution and use the `curve` function to overlay the normal curve. The function to be plotted by `curve` (in this case the `dnorm` normal density function) is always written as a function of x. Here x is a formal argument and does not reference a particular object in the workspace. Using the `add=TRUE` argument, we indicate that we wish to add this curve to the current plot and the `lwd` and `lty` arguments change the line thickness and line style of the displayed curve.

```
> mu = n * p; sigma = sqrt(n * p * (1 - p))
> curve(dnorm(x, mu, sigma), add=TRUE, lwd=2, lty=2)
```

To display the equation of the normal density on the graph, we can apply `text`, a general function for adding textual material to the current plot. The function `expression` can be used to add mathematical expressions. In the

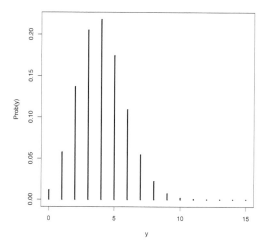

**Fig. 4.12** Display of the binomial probability distribution for $n = 20$ and $p = 0.2$.

following R code, we indicate that we wish to place the text at the location (10, 0.15) on the plot, and use **expression** to enter in the normal density formula. The syntax inside **expression** is similar to Latex syntax – **frac** indicates a fraction, **sqrt** indicates a square root, and **sigma** refers to the Greek letter $\sigma$. The **paste** function is used to concatenate two character strings.

```
> text(10, 0.15, expression(paste(frac(1, sigma*sqrt(2*pi)), " ",
+            e^{frac(-(y-mu)^2, 2*sigma^2)})), cex = 1.5)
```

A descriptive title is added to the graph using the **title** function.

```
> title("Binomial probs with n=2, p=0.2, and matching normal curve")
```

The manual page giving a more complete description of the syntax for **expression** is found by typing

```
?plotmath
```

Last, we wish to draw a line that connects the normal density equation to the curve. This is conveniently done using the **locator** function. If we use **locator(2)**, a cross-hair will show on the graph and we choose two locations on the grid by clicking the mouse. The x and y locations of the chosen points are stored in the list **locs**. The **arrows** function draws a line between the two locations on the current display; the resulting display is shown in Figure 4.13.

```
> locs = locator(2)
> arrows(locs$x[1], locs$y[1], locs$x[2], locs$y[2])
```

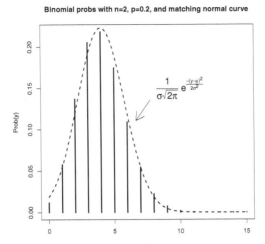

**Fig. 4.13** Display of the binomial probability distribution for $n = 20$ and $p = 0.2$.with text and title added.

## 4.10 Multiple Plots and Varying the Graphical Parameters

We use a snowfall data example to illustrate constructing multiple plots using the `layout` function and modifying some of the graphical parameters using the `par` function.

The website http://goldensnowball.blogspot.com describes the Golden Snowball Award given to the city in New York State receiving the most snowfall during the winter season. The variables `snow.yr1` and `snow.yr2` contain the 2009-10 and 2010-11 snowfalls (in inches through February 24, 2011) for the cities Syracuse, Rochester, Buffalo, Binghamton, and Albany.

```
> snow.yr1 = c(85.9, 71.4, 68.8, 58.8, 34.4)
> snow.yr2 = c(150.9, 102.0, 86.2, 80.1, 63.8)
```

We wish to construct a scatterplot and a difference plot in the same plotting window. We use the `windows` function to open a new graphics device of size 5 inches by 7 inches. The `layout` divides this device into two regions; the `matrix` argument divides the device into two regions of a single column and the relative heights of the two regions are 6 and 4.

```
> windows(width=5, height=7)
> layout(matrix(c(1, 2), ncol=1), heights=c(6, 4))
```

With the default settings, the *figure region* is the entire region of the graphics device and the *plot region* is the region where the data is graphed. The default location of the plot region can be displayed by showing the `plt` argument of `par`; the four values are represented in terms of fractions of the figure region. The region goes from 0.117 to 0.940 in the horizontal direction and 0.146 to 0.883 in the vertical direction.

```
> par("plt")
[1] 0.1171429 0.9400000 0.1459315 0.8826825
```

We wish to construct a scatterplot in the top region. We modify the `plt` and `xaxt` arguments to `par`. The values for `plt` result in a figure region with extra space on the right and less space below, and the `xaxt="n"` argument suppresses the plotting of the x axis. We use the `plot` function to construct a scatterplot of the two year snowfall amounts, use the `abline` function to overlay a line where the two snowfalls are equal, and label Syracuse's snowfall by the `text` function. The resulting display is shown in Figure 4.14.

```
> par(plt=c(0.20, 0.80, 0, 0.88), xaxt="n")
> plot(snow.yr1, snow.yr2, xlim=c(30, 100), ylim=c(30, 155),
+    ylab="2010-11 Snowfall (in)", pch=19,
+    main="Snowfall in Five New York Cities")
> abline(a=0, b=1)
> text(80, 145, "Syracuse")
```

The snowfall is measured in inches and we want to also display centimeters on the right axis to better communicate this graph to a non-American audience. We first save information about the current y tick mark locations (stored in the `yaxp` argument) in the variable `tm`. We use this information to compute the tick mark locations. Then we use the `axis` function to draw a right vertical axis; note that we compute the centimeter values by multiplying the inch values by 2.54 and rounding to the nearest 10 to make the tick labels more readable. Last, we use the `mtext` function to write a label on this new axis.

```
> tm = par("yaxp")
> ticmarks = seq(tm[1], tm[2], length=tm[3]+1)
> axis(4, at=ticmarks,
+    labels=as.character(round(2.54 * ticmarks, -1)))
> mtext("2010-11 Snowfall (cm)", side=4, line=3)
```

We next construct a scatterplot of the increase in snowfall (from 2009-10 to 2010-11) in the bottom region. The plot region is modified (using the `plt` argument) so that it matches the first graph horizontally and there is little space between the two regions. The `xaxt="s"` argument turns on the plotting of the x-axis. We plot the increase in snowfall against the 2009-10 amount and again identify the point corresponding to Syracuse. A right vertical axis corresponding to centimeters is again constructed using the same commands as in the top region.

```
> par(plt=c(0.20, 0.80, 0.35, 0.95), xaxt="s")
> plot(snow.yr1, snow.yr2 - snow.yr1, xlim=c(30, 100),
```

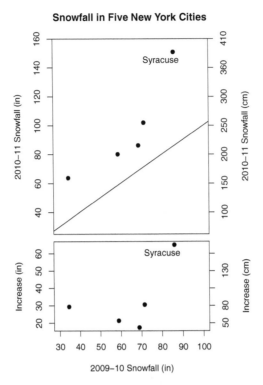

**Fig. 4.14** Two graphs to display the snowfall of five New York cities for two consecutive seasons. This display uses the `layout` function to divide the graphics device into unequal areas, the `plt` argument to `par` adjusts the figure regions, and the `axis` function is used to add a second axis to the plots.

```
+   xlab="2009-10 Snowfall (in)", pch=19,
+   ylab="Increase (in)")
> text(80, 60, "Syracuse")
> tm=par("yaxp")
> ticmarks=seq(tm[1], tm[2], length=tm[3] + 1)
> axis(4, at=ticmarks,
+     labels=as.character(round(2.54 * ticmarks, -1)))
> mtext("Increase (cm)", side=4, line=3)
```

The message from this plot is that all five New York cities had more snow during the 2010-11 season. The bottom plot focuses on the change in snowfall; Syracuse had over 60 inches more snowfall in 2010-11 and the remaining cities had an additional 25 inches of snow.

## 4.11 Creating a Plot using Low-Level Functions

Many graphics commands in R are high-level functions that will take care of all aspects of graph construction including drawing the axes, plotting the tick marks and tick labels, and choosing the points and lines to be displayed. In some situations one wishes to create a special-purpose graph with finer control over the construction process. Here we illustrate a simple example of constructing a graph from scratch.

*Example 4.4 (Graphing a circle with labels).*
   In Chapter 13, in our introduction of Markov Chains, in Figure 13.5, we constructed a graphical display of a circle on which six locations, numbered 1 through 6, are displayed. Since we don't want the plot axes to show, this is a good opportunity to illustrate some of the lower-level plotting functions in R.
   To begin, one needs to open a new graphics frame – this is accomplished by the `plot.new` function.

```
> plot.new()
```

A coordinate system is set up using the `plot.window` function. The minimum and maximum values of the horizontal and vertical scales are controlled by the `xlim` and `ylim` arguments . The `pty` argument indicates the type of plot region – the choice `pty = "s"` generates a square plotting region.

```
> plot.window(xlim=c(-1.5, 1.5), ylim=c(-1.5, 1.5), pty="s")
```

Now that the coordinate system has been set up, it is convenient to use polar coordinates to graph a circle. One defines a vector of angles `theta` equally spaced from 0 to $2\pi$ and the coordinates of the circle are the cosine and sine functions evaluated at these angles. The function `lines` with arguments x and y adds a line graph to the current graph. (The work is displayed in Figure 4.15.)

```
> theta = seq(0, 2*pi, length=100)
> lines(cos(theta), sin(theta))
```

Next, suppose one wishes to draw large solid plotting points at six equally-spaced locations on the circle. Six angle values are defined and stored in the vector `theta`. The function `points` adds points to the current graph. The arguments `cex = 3` plots the symbols at three times the usual size, and `pch = 19` uses a solid dot plotting symbol.

```
> theta=seq(0, 2*pi, length=7)[-7]
> points(cos(theta), sin(theta), cex=3, pch=19)
```

The points will be labelled outside of the circle using large numbers 1 through 6. To get precise control over the plotting locations, the `locator` function graphically select the locations. The `text` function places the `labels` one through six at these locations and the labels are drawn at two and one half times the usual size by specifying `cex = 2.5`. The plot is completed by drawing a box around the display with the `box` function.

```
> pos = locator(6)
> text(pos, labels=1:6, cex=2.5)
> box()
```

Figure 4.15 shows the completed graph. Using these low-level plot functions,

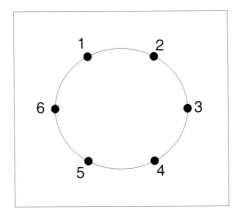

**Fig. 4.15** Graphical display of a circle with labeled locations constructed using low-level graphics functions in R.

the user has much control over the axes display and characteristics of the drawn lines and points. Using these tools, it is possible in principle to create many new types of graphical displays.

## 4.12 Exporting a Graph to a Graphics File

After creating a graph with R, one often wishes to export the graph to a graphics format such as *pdf*, *gif*, or *jpeg*, so the graph can be included in a document or web page. There are several ways of exporting graphics that we will outline through an example.

*Example 4.5 (Home run hitting in baseball history (continued)).*
    Suppose we wish to export the R graph in Figure 4.8 that displays the home run rates together with three lowess fits. In the Windows system, one

can simply save graphs using menu commands. After the graph has been constructed, select the graph and select the menu option "Save as" under the File menu. One selects the graphics format (metafile, postscript, pdf, png, bmp, or jpg) of interest and gives a name to the graphics file.

Alternatively, one can use functions to export graphs. For example, suppose we wish to save the same graph in the current working directory in a pdf format. In the R code, one first uses the function pdf to open the graphics device. The argument is the name of the saved graphics file, although one does not need to specify a name; the default name for saved files is "Rplots.pdf" that will overwrite any existing file with the same name. One next includes all of the functions for producing the graph. Then the dev.off function will close the graphics device and save the file.

```
> pdf("homerun.pdf")
> plot(Year, HR, xlab="Season",
+   ylab="Avg. Home Runs Hit by a Team in a Game",
+   main="Home Run Hitting in the MLB Across Seasons")
> lines(lowess(Year, HR))
> lines(lowess(Year, HR, f=1 / 3), lty=2)
> lines(lowess(Year, HR, f=1 / 6), lty=3)
> legend("topleft", legend=c("f = 2/3", "f = 1/3", "f = 1/6"), lty=c(1, 2, 3))
> dev.off()
```

$R_x$ **4.2** *It is a common mistake to forget to type the* **dev.off()** *function at the end. If this function is not included, the graphics file will not be created.*

There are similar functions such as postscript and jpeg for saving the graph using alternative graphics formats. See ?Devices for more details.

## 4.13 The lattice Package

Almost all of the graphs illustrated in this book are based on the traditional graphics system provided in the graphics package in R. Although this package provides much flexibility in producing a variety of statistical graphics, there are some alternative graphics systems available that extend and improve the traditional system. In this section we give a brief overview of the graphics available in the lattice package developed by Sarkar [43].

The graphics package contains a number of useful functions such as plot, barplot, hist, and boxplot for constructing statistical graphs. The lattice package contains a collection of plotting functions such as xyplot, barchart, histogram, and bwplot that produce similar types of graphs. By typing

```
> help(package=lattice)
```

one sees a display of functions and objects in the lattice package. Why should one use lattice functions instead of the traditional R graphing functions? First, there are several graphing functions in lattice that are not available in the traditional system. Second, the default appearance of some of the lattice graphs is sometimes better than the default appearance of the traditional graphs. Last, the lattice package provides some attractive extensions of the basic R graphics system and these extensions will be the focus of this section.

*Example 4.6 (Gas consumption of cars).*

To illustrate the lattice graphics functions, consider the dataset mtcars which contains the fuel consumption and other measures of design and performance for a group of 32 cars. Suppose one wishes to construct a scatterplot of miles per gallon against the weight of the car (lbs divided by 1000). In the lattice package, the basic syntax of the scatterplot function is

```
xyplot(yvar ~ xvar, data)
```

where data is the data frame containing the two variables. In the R script, the lattice package is loaded (using the library function) and a scatterplot of mpg and wt is constructed. Axis labels and a title are added by the same arguments as the traditional graphics system. The resulting graph is in Figure 4.16.

```
> library(lattice)
> xyplot(mpg ~ wt, data=mtcars, xlab="Weight",
+    ylab="Mileage",
+    main="Scatterplot of Weight and Mileage for 32 Cars")
```

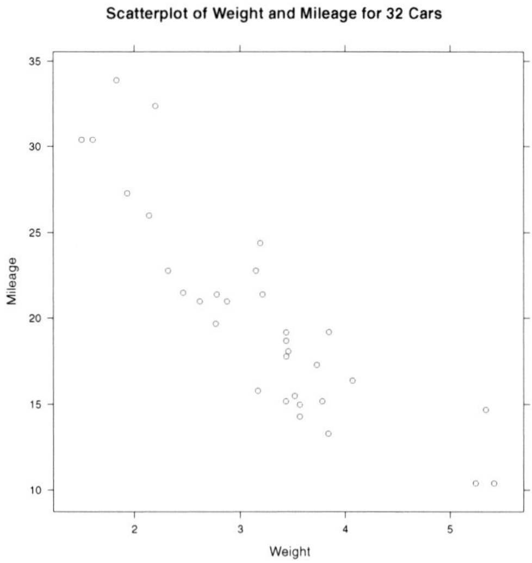

**Fig. 4.16** Scatterplot of weight and mileage for a sample of cars using the lattice xyplot function.

From the scatterplot one understands that the mileage of a car depends on the number of cylinders in the engine. If one controls for the number of cylinders, is there still an association between mileage and weight? One nice extension offered by the `lattice` package is the availability of *conditional* graphs, that is, graphs conditional on the values of a third variable. In the case of a scatterplot, this conditional plot has the basic syntax

```
xyplot(yvar ~ xvar | cvar, data)
```

where `cvar` is a third variable with a limited number of values. In this example, the variable `cyl` is number of cylinders in the engine, and `xyplot` function is used to construct a separate scatterplot of mileage and weight for each possible number of cylinders. To help in visualizing the plotting symbols in Figure 4.17, solid plotting symbols (`pch = 19`) are drawn one and a half times larger than the usual size (`cex = 1.5`).

```
> xyplot(mpg ~ wt | cyl, data=mtcars, pch=19, cex=1.5,
+    xlab="Weight", ylab="Mileage")
```

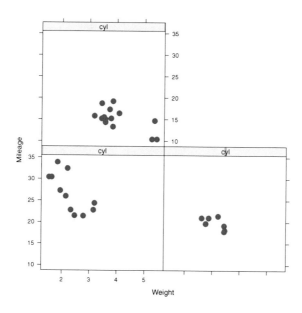

**Fig. 4.17** Scatterplot of weight and mileage conditional on the number of cylinders using the `lattice xyplot` function.

Since the scaling is the same for all three plots, one can easily compare the three scatterplots. The panels in the lower left, lower right, and upper left sections of the figure correspond respectively to four, six, and eight cylinders.

For the 4-cylinder cars, there seems to be a relatively strong relationship between mileage and weight. In contrast, for 8-cylinder cars, the relationship between mileage and weight seems weaker.

The `lattice` package also allows for graphical displays that allow for comparisons between different data groups. Suppose that one is interested in comparing the weights of 4-cylinder, 6-cylinder, and 8-cylinder cars. A density plot is an estimate of the smooth distribution of a variable. One can construct density plots of different groups of data using the syntax

```
> densityplot(~ yvar, group=gvar, data)
```

where `yvar` is the continuous variable of interest, and `gvar` is the grouping variable. This type of graph can be used to compare the distributions of weights of the three types of cars. (See Figure 4.18.) The `auto.key` option will add a legend above the graph showing the line color of each group.

```
> densityplot(~ wt, groups=cyl, data=mtcars,
+     auto.key=list(space="top"))
```

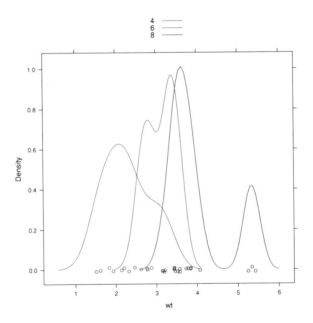

**Fig. 4.18** Density plots of car weights for 4-cylinder, 6-cylinder, and 8-cylinder cars using the `lattice` `densityplot` function with the `groups` argument option.

Looking at this graph, one sees the variability of weights for each group of cars. Four-cylinder cars tend to weigh between 1000-4000 pounds and six-cylinder cars between 2000-4000 pounds. The eight-cylinder cars appear to cluster in two groups, one group weighing 3000-4000 pounds and a second group between 5000-6000 pounds.

*Example 4.7 (Sample means).*

In Chapter 2, we considered collections of triples of random numbers generated using the the RANDU algorithm. Although this algorithm is supposed to produce uncorrelated triples, it has been well-documented that the simulated numbers have an association pattern. This can be easily shown using a a three-dimensional scatterplot drawn by the `lattice cloud` function.

```
> library(lattice)
> cloud(z ~ x + y, data = randu)
```

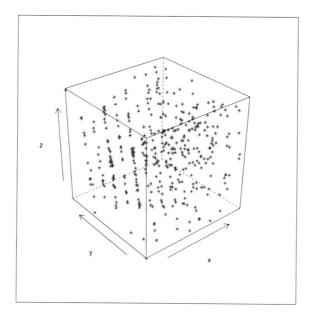

**Fig. 4.19** Cloud plot of RANDU random number triples using the `lattice` package.

Looking carefully at Figure 4.19, one sees groups of points $(x, y, z)$ that follow linear patterns of the type

$$z = ax + by$$

for constants $a$ and $b$. The algorithm RANDU was used as the random number generation algorithm for the early personal computers, but currently alternative algorithms (without this dependence problem) are used for simulating random numbers in modern computers.

## 4.14 The ggplot2 Package

A second alternative to traditional graphics in R is the ggplot2 package
described in Wickham.[52] This system is based on a different way of thinking
about graphics. Specifically, ggplot2 is an R implementation of the "grammar
of graphics" described in Wilkinson [55]. This alternative graphics system is
described in the context of an interesting sports example.

*Example 4.8 (World records in the mile run ).*
  Professional athletes are continuously setting new world records in many
sports such as running, jumping, and swimming, and it is interesting to explore the patterns of these world records over time. The data set
"world.record.mile.csv" contains the world record times in the mile run for
men and for women. Obviously, the world record times have decreased over
time, but we are interested in learning about the pattern of decrease, and
seeing if the pattern of decrease is different for men and women.
  We read in the dataset using the read.csv function. We are interested in
only the record times that were obtained in the year 1950 or latter; we use
the subset function to create a new data frame mile2 containing the times
for these recent world records.

```
> mile = read.csv("world.record.mile.csv")
> mile2 = subset(mile, Year >= 1950)
```

Wilkinson [55] describes a statistical graphic as a combination of independent components. Variables in a dataset are assigned particular roles or
*aesthetics*; the aesthetic for one variable might be the plotting position along
the horizontal axis and the aesthetic for a second variable might be the color
or shape of the plotting point. Once the aesthetics for a set of variables are
defined, then one uses a *geometric object* or *geom* to construct a graph. Examples of geoms are points, lines, bars, histograms, and boxplots.
  One thinks of constructing a graph by overlaying a set of layers on a grid.
To construct a scatterplot, one overlays a set of point geoms on a Cartesian
coordinate system. If one is interested in summarizing the pattern of points,
then one might wish to overlay a smoothing curve on the current graph. A
lowess smoother is an example of a *statistic*, a transformation of the data
that you wish to graph.
  There are many choices how the variables are mapped to the aesthetic
properties. For example, variables may be graphed on a log scale, or one
wishes to map a variable using a specific range of colors. These mappings are
called *scales* and there are many choices for scales in the ggplot2 package.
The graphics system also allows for *position adjustments*, fine tune positioning
of graphical objects such as jittering and stacking, and *faceting*, plotting
subsets of the data on different panels.
  For our example, we first load in the ggplot2 package. The ggplot function is used to initialize the graph – the first argument is the name of the

data frame `mile2` and the second argument `ase()` defines the aesthetics that map variables in the data frame to aspects of the graph. We indicate in the `aes` argument that `Year` is to be plotted along the horizontal (x) axis, seconds is to be plotted along the vertical (y) axis, and the variable `Gender` will be used as the color and shape aesthetics. We store this graph initialization information in the variable `p`.

```
> library(ggplot2)
> p = ggplot(mile2, aes(x = Year, y = seconds,
+   color = Gender, shape = Gender))
```

The function `ggplot` does not perform any plotting – it just defines the data frame and the aesthetics. We construct a plot by adding geom and statistic layers to this initial definition. A scatterplot is constructed using the `geom_point` function; the `size = 4` arguments indicates that the points will be drawn twice the usual size of 2. Then we add a smoothing lowess curve using the `geom_smooth` function.

```
> p + geom_point(size = 4) + geom_smooth()
```

Figure 4.20 displays the resulting graphical display. Since the variable **Gender** has been assigned to the color and shape aesthetics, note that the male and female record times are plotted using different colors and shapes and a legend is automatically drawn outside of the plotting region. Note that since `Gender` has a color aesthetic, the smoothing curves are graphed for each gender. The record times appear to be linearly decreasing as a function of year for both genders this time period, although the rate of decrease is much greater for the women times.

Although the "grammar" of the `ggplot2` system may seem odd at first look, one can construct attractive and useful graphics using a limited amount of R code. The book Hadley [52] provides a comprehensive description of the `ggplot2` package and many graphics examples are displayed on the accompanying website `had.co.nz/ggplot2`.

## Exercises

**4.1 (Speed and stopping distance).** The data frame `cars` in the `datasets` package gives the speed (in mph) and stopping distance (in ft) for 50 cars.

a. Use the `plot` function to construct a scatterplot of `speed` (horizontal) against `dist` (vertical).
b. Revise the basic plot by labeling the horizontal axis with "Speed (mpg)" and the vertical axis with "Stopping Distance (ft)," Add a meaningful title to the plot.
c. Revise the plot by changing the plot symbol from the default open circles to red filled triangles (`col="red"`, `pch=17`).

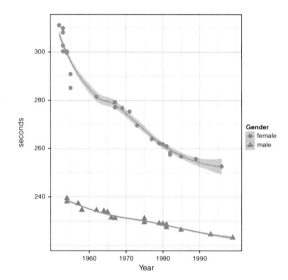

**Fig. 4.20** Scatterplots of world record running times for the mile for the men and women using the `ggplot2` package. Smoothing lines are added to show the general pattern of decrease for each gender.

**4.2 (Speed and stopping distance (continued)).** Suppose that one wishes to compare linear and quadratic fits to the (`speed, dist`) observations. One can construct these two fits in R using the code

```
fit.linear = lm(dist ~ speed, data=cars)
fit.quadratic = lm(dist ~ speed + I(speed^2), data=cars)
```

a. Construct a scatterplot of speed and stopping distance.
b. Using the `abline` function with argument `fit.linear`, overlay the best line fit using line type "dotted" and using a line width of 2.
c. Using the `lines` function, overlay the quadratic fit using line type "long-dash" and a line width of 2.
d. Use a legend to show the line types of the linear and quadratic fits.
e. Redo parts (a) - (d) using two contrasting colors (say red and blue) for the two different fits.

**4.3 (Speed and stopping distance (continued)).**

a. Construct a residual plot for the linear fit by typing

```
plot(cars$speed, fit.linear$residual)
```

b. Add a blue, thick (`lwd=3`) horizontal line to the residual plot using the `abline` function.
c. There are two large positive residuals in this graph. By two applications of the `text` function, label each residual using the label "POS" in blue.

d. Label the one large negative residual in the graph with the label "NEG" in red.

e. Use the `identify` function to find the row numbers of two observations that have residuals close to zero.

**4.4 (Multiple graphs).** The dataset `mtcars` contains measurements of fuel consumption (variable `mpg`) and other design and performance characteristics for a group of 32 automobiles. Using the `mfrow` argument to `par`, construct scatterplots of each of the four variables `disp` (displacement), `wt` (weight), `hp` (horsepower), `drat` (rear axle ratio) with mileage (`mpg`) on a two by two array. Comparing the four graphs, which variable among displacement, weight, horsepower, and rear axle ratio has the strongest relationship with mileage?

**4.5 (Drawing houses).** The following function `house` plots an outline of a house centered about the point (`x`, `y`):

```
house=function(x, y, ...){
  lines(c(x - 1, x + 1, x + 1, x - 1, x - 1),
    c(y - 1, y - 1, y + 1, y + 1, y - 1), ...)
  lines(c(x - 1, x, x + 1), c(y + 1, y + 2, y + 1), ...)
  lines(c(x - 0.3, x + 0.3, x + 0.3, x - 0.3, x - 0.3),
    c(y - 1, y - 1, y + 0.4, y + 0.4, y - 1), ...)
}
```

a. Read the function `house` into R.

b. Use the `plot.new` function to open a new plot window. Using the `plot.window` function, set up a coordinate system where the horizontal and vertical scales both range from 0 to 10.

c. Using three applications of the function `house`, draw three houses on the current plot window centered at the locations (1, 1), (4, 2), and (7, 6).

d. Using the ... argument, one is able to pass along parameters that modify attributes of the `line` function. For example, if one was interested in drawing a red house using thick lines at the location (2, 7), one can type

```
house(2, 7, col="red", lwd=3)
```

Using the `col` and `lty` arguments, draw three additional houses on the current plot window at different locations, colors, and line types.

e. Draw a boundary box about the current plot window using the `box` function.

**4.6 (Drawing beta density curves).** Suppose one is interesting in displaying three members of the beta family of curves, where the beta density with shape parameters $a$ and $b$ (denoted by Beta($a,b$)) is given by

$$f(y) = \frac{1}{B(a,b)} y^{a-1}(1-y)^{b-1}, \, 0 < y < 1.$$

One can draw a single beta density, say with shape parameters $a = 5$ and $b = 2$, using the `curve` function:

```
curve(dbeta(x, 5, 2), from=0, to=1))
```

a. Use three applications of the `curve` function to display the Beta(2, 6), Beta(4, 4), and Beta(6, 2) densities on the same plot. (The `curve` function with the `add=TRUE` argument will add the curve to the current plot.)
b. Use the following R command to title the plot with the equation of the beta density.

```
title(expression(f(y)==frac(1,B(a,b))*y^{a-1}*(1-y)^{b-1}))
```

c. Using the `text` function, label each of the beta curves with the corresponding values of the shape parameters $a$ and $b$.
d. Redraw the graph using different colors or line types for the three beta density curves.
e. Instead of using the `text` function, add a legend to the graph that shows the color or line type for each of the beta density curves.

**4.7 ( `lattice` graphics).** The dataset `faithful` contains the duration of the eruption (in minutes) `eruptions` and the waiting time until the next eruption `waiting` (in minutes) for the Old Faithful geyser. One is interested in exploring the relationship between the two variables.

a. Create a factor variable `length` that is "short" if the eruption is smaller than 3.2 minutes, and "long" otherwise.

```
faithful$length = ifelse(faithful$eruptions < 3.2,
  "short", "long")
```

b. Using the `bwplot` function in the `lattice` package, construct parallel boxplots of the waiting times for the "short" and "long" eruptions.
c. Using the `densityplot` function, construct overlapping density plots of the waiting times of the "short" and "long" eruptions.

**4.8 ( `ggplot2` graphics).** In Exercise 4.7, the waiting times for the Old Faithful geysers were compared for the short and long eruptions where the variable `length` in the `faithful` data frame defines the duration of the eruption.

a. Suppose a data frame `dframe` contains a numeric variable `num.var` and a factor `factor.var`. After the `ggplot2` package has been loaded, then the R commands

```
ggplot(dframe, aes(x = num.var, color = factor.var))
  + geom_density()
```

will construct overlapping density estimates of the variable `num.var` for each value of the factor `factor.var`. Use these commands to construct overlapping density estimates of the waiting times of the geysers with short and long eruptions.
b. With a data frame `dframe` containing a numeric variable `num.var` and a factor `factor.var`, the `ggplot2` syntax

```
ggplot(dframe, aes(y = num.var, x = factor.var))
  + geom_boxplot()
```

will construct parallel boxplots of the variable num.var for each value of the factor factor.var. Use these commands to construct parallel boxplots of the waiting times of the geysers with short and long eruptions.

# Chapter 5
# Exploratory Data Analysis

## 5.1 Introduction

Exploratory data analysis is the process by which a person manipulates data with the goal of learning about general patterns or tendencies and finding specific occurrences that deviate from the general patterns. Much like a detective explores a crime scene, collects evidence and draws conclusions, a statistician explores data using graphical displays and suitable summaries to draw conclusions about the main message of the data.

John Tukey and other statisticians have devised a collection of methods helpful in exploring data. Although the specific data analysis techniques are useful, exploratory data analysis is more than the methods – it represents an attitude or philosophy about how data should be explored. Tukey makes a clear distinction between *confirmatory* data analysis, where one is primarily interested in drawing inferential conclusions, and *exploratory* methods, where one is placing few assumptions on the distributional shape of the data and simply looking for interesting patterns. Good references on exploratory methods are Tukey [47] and Hoaglin et al. [22].

There are four general themes of exploratory data analysis, namely *Revelation, Resistance, Residuals,* and *Reexpression,* collectively called the *four R's.* There is a focus on *revelation,* the use of suitable graphical displays in looking for patterns in data. It is desirable to use *resistant* methods – these methods are relatively insensitive to extreme observations that deviate from the general patterns. When we fit simple models such as a line, often the main message is not the fitted line, but rather the *residuals,* the deviations of the data from the line. By looking at residuals, we often learn about data patterns that are difficult to see by the initial data displays. Last, in many situations, it can be difficult to see patterns due to the particular measuring scale of the data. Often there is a need to *reexpress* or change the scale of the data. Well-chosen reexpressions, such as a log or square root, make it easier to see general patterns and find suitable data summaries. In the following example, we illustrate each of the four "R themes" in exploratory work.

## 5.2 Meet the Data

*Example 5.1 (Ratings of colleges).*
    It can be difficult for an American high school student to choose a college. To help in this college selection process, *U.S. News and World Report* (http://www.usnews.com) prepares a yearly guide *America's Best Colleges*. The 2009 guide ranks all of the colleges in the United States with respect to a number of different criteria. The dataset `college.txt` contains data in the guide collected from a group of "National Universities." These are schools in the United States that offer a range of degrees both at the undergraduate and graduate levels. The following variables are collected for each college:

a. School – the name of the college
b. Tier – the rank of the college into one of four tiers
c. Retention – the percentage of freshmen who return to the school the following year
d. Grad.rate – the percentage of freshman who graduate in a period of six years
e. Pct.20 – the percentage of classes with 20 or fewer students
f. Pct.50 – the percentage of classes with 50 or more students
g. Full.time – the percentage of faculty who are hired full-time
h. Top.10 – the percentage of incoming students who were in the top ten percent of their high school class
i. Accept.rate – the acceptance rate of students who apply to the college
j. Alumni.giving – the percentage of alumni from the college who contribute financially

We begin by loading the dataset into R using the function `read.table` and saving it in the data frame `dat`. Note that the `sep` argument indicates there are tabs separating the columns in the data file.

```
> dat = read.table("college.txt", header=TRUE, sep="\t")
```

There are some colleges where not all of the data were collected. The R function `complete.cases` will identify the colleges where all of the variables have been collected and the `subset` function is used to create a new data frame `college` containing only the colleges with "complete" data.

```
> college = subset(dat, complete.cases(dat))
```

## 5.3 Comparing Distributions

One measure of the quality of a college is the variable `Retention`, the percentage of freshmen who return to the college the following year. We wish to display the distribution of the retention rates and compare the distribution of rates across different college subgroups.

### 5.3.1 Stripcharts

One basic graph of the retention rates is a stripchart or one-dimensional scatterplot constructed using the `stripchart` function. Using the `method = "stack"` option, the dots in the graph will be stacked, and the option `pch = 19` will use solid dots as the plotting character.

```
> stripchart(college$Retention, method="stack", pch=19,
+      xlab="Retention Percentage")
```

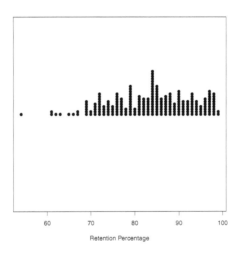

**Fig. 5.1** Stripchart of the retention percentages of all National universities.

From Figure 5.1, we see much variability in the retention rates from 55% to near 100% and wonder which variables are helpful in explaining this variation. One of the general measures of a school's quality is its Tier (either 1, 2, 3, or 4) and a next step might be to construct parallel stripcharts of the retention rates by Tier. This graph is constructed by a slight variation of stripchart – the argument `Retention ~ Tier` indicates that we wish separate displays of retention for each of the four tiers. Note that we don't use the `college$Retention` syntax, since we have indicated by the `data=college` argument that the data frame `college` is being used.

```
> stripchart(Retention ~ Tier, method="stack", pch=19,
+     xlab="Retention Percentage",
+     ylab="Tier", xlim=c(50, 100), data=college)
```

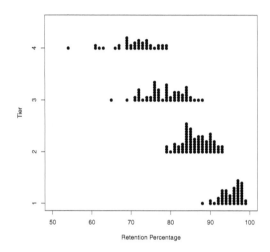

**Fig. 5.2** Parallel stripcharts of the retention percentages of all National universities grouped by tier.

It is clear from Figure 5.2 that the retention percentage differs among the four groups of colleges. For the Tier 1 schools, the percentages primarily fall between 90 and 100; in contrast, most of the retention percentages of the Tier 4 schools fall between 65 and 80.

## 5.3.2 Identifying outliers

The parallel stripcharts are helpful in seeing the general distribution of retention percentages for each tier. Also from the graph, we notice a few schools with retention percentages that seem set apart from the other colleges in the same tier. The `identify` function is helpful in identifying the schools with these unusual percentages. In this function, we give the $x$ and $y$ variables of the plot, indicate by the `n=2` option that we wish to identify two points, and the `labels=college$School` option indicates that we wish to label the points by the school names. When this function is executed, a cross-hair will appear over the graph and one moves the mouse and clicks at the locations of the two outliers. At each click, the name of the school will appear next to the plotting point. In the R console window, the row numbers of the data frame corresponding to these two schools are displayed.

```
> identify(college$Retention, college$Tier, n=2,
+   labels=college$School)
[1] 158 211
```

We see in Figure 5.3 that the two schools with unusually small retention percentages (relative to the other schools in the same tier) are Bridgeport in Tier 4 and South Carolina State in Tier 3.

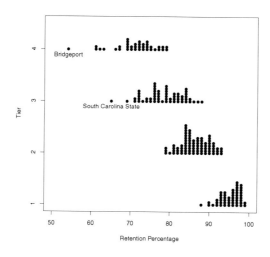

**Fig. 5.3** Parallel stripcharts of the retention percentages of all National universities grouped by tier. Two outliers are identified by the school name.

## 5.3.3 Five-number summaries and boxplots

The parallel stripchart display shows differences in the retention percentages among the four tiers of colleges, and a next step is to summarize these differences. A good collection of summaries of a dataset is the median, the lower and upper quartiles, and the low and high values. This group of summaries is called (for obvious reasons) the *five-number summary* and a *boxplot* is a graph of these five numbers. The boxplot function will compute five-number summaries for each group and display parallel boxplots (see Figure 5.4). As in the `stripchart` function, the `Retention` ∼ `Tier` formula indicates that we wish to construct boxplots of retention by tier, and the `horizontal=TRUE` argument indicates the boxplots will be displayed in a horizontal style. As in the stripchart function, the `data=college` argument indicates that the variables are part of the data frame `college`. The resulting boxplot display is shown in Figure 5.4. The locations of the median retention values for the four tiers are shown by the dark lines in the boxes, and the spreads of the four graphs are reflected by the widths of the boxes. Using an EDA rule for flagging outliers, the display shows four schools (indicated by separate points) whose retention percentages are unusually small for their associated tiers.

```
> b.output = boxplot(Retention ~ Tier, data=college, horizontal=TRUE,
+    ylab="Tier", xlab="Retention")
```

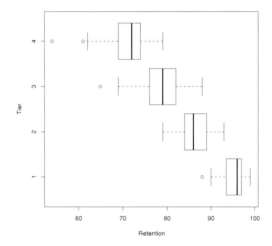

**Fig. 5.4** Boxplots of the National university retention percentages grouped by Tier.

The output of `boxplot` has been saved to the variable `b.output`, a list with the different components of the boxplot computations. One can display the five-number summaries by the `stats` component of `b.output`:

```
> b.output$stats
       [,1] [,2] [,3] [,4]
[1,] 90.0   79   69   62
[2,] 93.5   84   76   69
[3,] 96.0   86   79   72
[4,] 97.0   89   82   74
[5,] 99.0   93   88   79
attr(,"class")
        1
"integer"
```

A column of `b.output$stats` corresponds to the five-number summary of the retention rates for a particular tier. We see the five-number summaries of tiers 3 and 4 are respectively (69, 76, 79, 82, 88) and (62, 69, 72, 74, 79). One can measure the spread of the two groups of data by the quartile spread, the distance between the two quartiles. The quartile spread of the retentions for Tier 3 schools is $82 - 76 = 6$ and the quartile spread of the Tier 4 schools is $74 - 69 = 5$. Since the spreads of the two groups are similar, one can compare the medians – the median for the Tier 3 retentions is 79 and the median for the Tier 4 retentions is 72. We observe that the retention percentages tend to be $79 - 72 = 7$ points higher for Tier 3 than for Tier 4. In addition to the five-number summaries, information about the outliers is also stored. The `out` and `group` components of `b.output` give the outlying values and their corresponding groups.

```
> b.output$out
[1] 88 65 61 61 54

> b.output$group
[1] 1 3 4 4 4
```

From Figure 5.4, there were two visible outliers in tier 4, but the output indicates that there are actually three outliers in this tier, corresponding to retention percentages of 54, 61, and 61.

## 5.4 Relationships Between Variables

### 5.4.1 Scatterplot and a resistant line

Since it is reasonable to believe a school's first-year retention percentage will affect its graduation percentage, we next look at the relationship between the variables `Retention` and `Grad.rate`. Using the `plot` function, we construct a scatterplot and the resulting graph is shown in Figure 5.5. As expected, we see a strong positive association between first-year retention rate and the graduation rate.

```
> plot(college$Retention, college$Grad.rate,
+   xlab="Retention", ylab="Graduation Rate")
```

For exploratory work, it is useful to fit a line that is resistant or not sensitive to outlying points. One fitting method of this type is Tukey's "resistant line" implemented by the function `line`. Essentially, the resistant line procedure divides the scatterplot into left, middle, and right regions, computes resistant summary points for each region, and finds a line from the summary points. We fit this resistant line to these data and the fitting calculations are stored in the variable `fit`. In particular, the coefficients of the fitted line are stored in `fit$coef`:

```
> fit = line(college$Retention, college$Grad.rate)
> coef(fit)
[1] -83.631579   1.789474
```

The fitted line is given by

$$\text{Graduation Rate} = -83.63 + 1.79 \times \text{Retention Rate}.$$

The slope of this line is 1.79 – for every one percent increase in the retention rate, the average graduation rate increases by 1.79%. The line on the scatterplot is added by the **abline** function in Figure 5.5.

```
> abline(coef(fit))
```

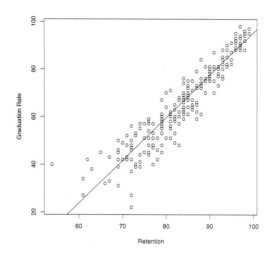

**Fig. 5.5** scatterplot of the graduation rates and retention rates for the National universities. A resistant best-fitting line is placed on top of the graph.

## 5.4.2 Plotting residuals and identifying outliers

In exploratory work, we wish to look beyond the obvious relationship between retention and graduation rates by examining schools deviating from the general straight-line pattern. We look further by considering the *residuals*, the differences between the actual graduation rates and the values predicted from the fitted resistant line. The set of residuals are stored in the list element **fit$residuals** and the **plot** function is used to construct a scatterplot of residuals against the retention rates in Figure 5.6. A horizontal line at zero is added using **abline** to help in interpreting this plot. Positive residual values correspond to observed graduation rates that are larger than the values predicted from the straight-line relation, and negative residuals correspond to graduate rates smaller than the predicted rates.

```
> plot(college$Retention, fit$residuals,
+   xlab="Retention", ylab="Residual")
> abline(h=0)
```

We learn some new things from inspecting this residual plot. There is a fan-shaped appearance in the graph, indicating that the spread of the residuals is higher for low retention schools than for high retention schools. Most of the residuals fall between $-20$ and 20 percentage points, indicating that the observed and predicted graduation rates fall within 20 points for most schools. We do notice two unusually large residuals, and we identify and label these residuals using the **identify** function.

```
> identify(college$Retention, fit$residuals, n=2,
+   labels=college$School)
```

When this function is executed, a crosshair will appear on the graph. One clicks at the locations of the two large residuals and the names of the schools appear next to the plotting points. (See Figure 5.6.) The two large residuals correspond to the schools Bridgeport and New Orleans. Although Bridgeport has a relatively low retention percentage, it has a large positive residual which indicates that its graduation percentage is large given its retention percentage. Perhaps Bridgeport's actual retention percentage is higher than what was recorded in this dataset. In contrast, New Orleans has a large negative residual. This school's graduation percentage is lower than one would predict from its retention percentage. This suggests that another variable (a so-called *lurking variable*) may explain New Orleans' low graduation percentage.

## 5.5 Time Series Data

### *5.5.1 Scatterplot, least-squares line, and residuals*

*Example 5.2 (Enrollment growth at a university).*
    Bowling Green State University celebrated its centennial in 2010 and it published online its enrollment counts for the years 1914 through 2008. The dataset "bgsu.txt" contains the enrollment counts for the significant growth years 1955 through 1970. We read in the dataset and use the **plot** function to construct a scatterplot of **Enrollment** against **Year**. (See Figure 5.7.)

```
> bgsu = read.table("bgsu.txt", header=TRUE, sep="\t")
> plot(bgsu$Year, bgsu$Enrollment}
```

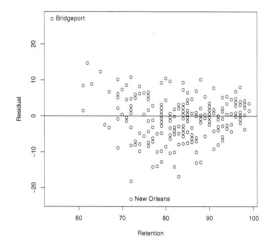

**Fig. 5.6** Plot of the residuals of a resistant fit to the graduation percentages by the retention percentages for the National universities. Two unusually large residuals are labeled with the corresponding college name.

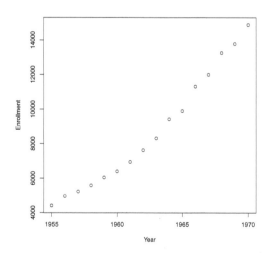

**Fig. 5.7** scatterplot of BGSU enrollment against year for the growth period 1955 to 1970.

To help us understand the pattern of growth of enrollment, we fit a line. The lm function is used to fit a least-squares line with the calculations stored in the variable fit. The fitted line is placed on the scatterplot using the abline function with the argument fit. The vector of residuals is stored in the component residuals of fit and the plot function is used to construct a scatterplot of the residuals against year. The abline function is used with the h=0 argument to add a horizontal line at zero to the residual plot. Figure 5.8 shows the two plots.

```
> fit = lm(Enrollment ~ Year, data=bgsu)
> abline(fit)
> plot(bgsu$Year, fit$residuals)
> abline(h=0)
```

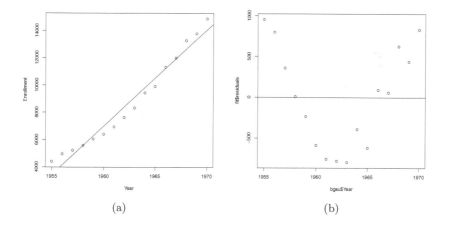

(a)                                        (b)

**Fig. 5.8** Least-squares fit to enrollment data (a) and residual plot (b). There is a clear curvature pattern to the residuals, indicating that the enrollment is not increasing in a linear fashion.

Looking at the residual graph, there is a clear curvature pattern in the residuals, indicating that BGSU's enrollment is not increasing in a linear way. An alternative model may better describe the enrollment growth.

### 5.5.2 Transforming by a logarithm and fitting a line

Suppose instead that the BGSU enrollment is increasing exponentially. This means that, for some constants $a$ and $b$, the enrollment follows the relationship

$$Enrollment = a \exp(bYear).$$

If we take the logarithm of both sides of the equation, we obtain the equivalent linear relationship between the log of enrollment and year

$$\log Enrollment = \log a + bYear.$$

We can find suitable constants $a$ and $b$ by fitting a line to the (Year, log Enrollment) data. In the following R code, we define a new variable `log.Enrollment` containing the log enrollment values.

```
> bgsu$log.Enrollment = log(bgsu$Enrollment)
```

$\mathbf{R_x}$ **5.1** *The syntax* `bgsu$log.Enrollment` *on the left side of the assignment creates a new variable* `log.Enrollment` *in the* `bgsu` *data frame. The logarithm of enrollment is then assigned to this new variable.*

We construct a scatterplot of the reexpressed data against year, and use the `lm` function to fit a line to these data. Figure 5.9 displays the least-squares fit to the log enrollment data, and plots the corresponding residuals. Generally, it seems that a linear pattern is a closer fit to the log enrollment than for the enrollment. Looking at the residual graph, there is a "high, low, high, low" pattern in the residuals as one looks from left from right, but we do not see the strong curvature pattern that we saw in the residuals from the linear fit to the enrollment.

```
> plot(bgsu$Year, bgsu$log.Enrollment)
> fit2 = lm(log.Enrollment ~ Year, data=bgsu)
> fit2$coef
  (Intercept)          Year
-153.25703366    0.08268126
> abline(fit2)
> plot(bgsu$Year, fit2$residuals)
> abline(h=0)
```

From the R output, we see that the least-squares fit to the log enrollment data is

$$\log Enrollment = -153.257 + 0.0827Year.$$

This is equivalent to the exponential fit

$$Enrollment = \exp(-153.257 + 0.0827Year) \propto (1.086)^{Year},$$

where $\propto$ means "is proportional to." We see that BGSU's enrollment was increasing approximately 8.6% a year during the period between 1955 and 1970.

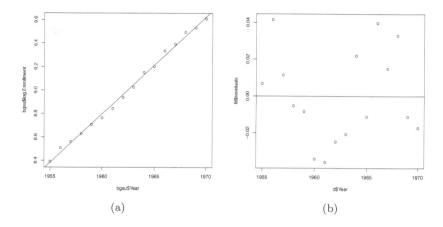

**Fig. 5.9** Least-squares fit to the log enrollment data (a) and the residual plot (b). Since there is no strong curvature pattern in the residuals, a linear fit seems more appropriate for the log enrollment than for the enrollment.

## 5.6 Exploring Fraction Data

### 5.6.1 Stemplot

*Example 5.3 (Ratings of colleges (continued)).*

One measure of quality of a university is the percentage of incoming students who graduated in the top ten percent of their high school class. Suppose we focus our attention at the "Top Ten" percentages for the Tier 1 colleges. We first use the **subset** function to extract the Tier 1 schools and put them in a new data frame **college1**:

```
> college1 = subset(college, Tier==1)
```

A stemplot of the percentages can be produced using the **stem** function.

```
> stem(college1$Top.10)
   4 | 3
   5 | 589
   6 | 344468
   7 | 355599
   8 | 02445556777888
   9 | 00223334566677777889
  10 | 0
```

## 5.6.2 Transforming fraction data

Since the percentages are left-skewed with a cluster of values in the 90's, it is a little difficult to summarize and hard to distinguish the schools with high percentages of "top ten" students. This suggests that we might be able to improve the display by an appropriate reexpression. For percentage data or equivalently fraction data that are piled up at the extreme values of 0 and 1, Tukey suggests the use of several reexpressions. The *folded fraction* is defined by

$$ff = f - (1 - f).$$

This reexpression expands the scale from the interval $(0, 1)$ to the interval $(-1, 1)$; a fraction $f = 0.5$ is a folded fraction of $ff = 0$. The *folded root* or *froot* is defined as

$$froot = \sqrt{f} - \sqrt{1 - f}$$

and the *folded log* or *flog* is defined as

$$flog = \log(f) - \log(1 - f).$$

Figure 5.10 displays the values of these reexpression for particular values of the fraction $f$. This figure was created using the following R code:

```
f = c(0.05, 0.1, 0.3, 0.5, 0.7, 0.9, 0.95)
ff = f - (1 - f)
froot = sqrt(2 * f) - sqrt(2 * (1 - f))
flog = 1.15 * log10(f) - 1.15 * log10(1 - f)
D = data.frame(f, ff, froot, flog)
matplot(t(as.matrix(D)), 1:4, type="l", lty=1, lwd=1,
  xlab="FRACTION", ylab="TRANSFORMATION",
  xlim=c(-1.8, 2), ylim=c(0.5, 4.3))
matplot(t(as.matrix(D[c(1, 4, 7), ])),
  1:4, type="l", lwd=3, lty=1, add=TRUE)
lines(D[c(1, 7), 1], c(1, 1), lwd=2)
lines(D[c(1, 7) ,2], 2 * c(1, 1), lwd=2)
lines(D[c(1, 7), 3], 3 * c(1, 1), lwd=2)
lines(D[c(1, 7), 4], 4 * c(1, 1), lwd=2)
text(c(1.8, 1.5, 1.3, 1.3, 0, 0.5 ,1),
  c(4, 3, 2, 1, 0.8, 0.8, 0.8),
  c("flog", "froot", "ff", "f", "f=.05", "f=.5", "f=.95"))
```

Figure 5.10 illustrates several desirable properties of these reexpressions. First, they are symmetric reexpressions in the sense that the ff or froot or flog of $f$ will be the negative of the ff or froot or flog of $1 - f$. Also the froot and flog rexpressions have the effect of expanding the scale for fractions close to 0 or 1.

We compute these reexpressions of the Top Ten percentages. To avoid problems with computing logs at percentages of 0 and 100, a value of 0.5 is added to the percentages of Top Ten and "not Top Ten" before the flog is taken.

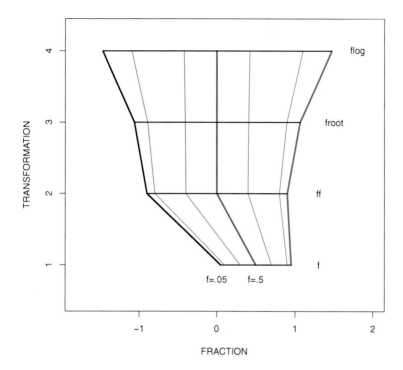

**Fig. 5.10** Display of three different reexpressions for fraction data. The bottom line (labelled f) displays fraction values of 0.05, 0.10, 0.30, 0.50, 0.70, 0.90, 0.95, and the ff, froot, and flog lines display the fractions on the folded fraction, folded root, and folded log scales.

```
> froot = sqrt(college1$Top.10) - sqrt(100 - college1$Top.10)
> flog = log(college1$Top.10 + 0.5) - log(100 - college1$Top.10 + 0.5)
```

Stemplots of the froot and flog Top 10 percentages are displayed. Both reexpressions have the effect of making the Top 10 percentages more symmetric and spreading out the schools with high values.

```
> stem(froot}
  The decimal point is at the |

  -0 | 0
   0 | 7139
   2 | 000363777
   4 | 3358223335777999
   6 | 338800025888
   8 | 11111559
  10 | 0
```

```
> stem(flog)
  The decimal point is at the |

  -0 | 3
   0 | 234566677
   1 | 01113345667778999
   2 | 000224455579
   3 | 1113333377
   4 | 2
   5 | 3
```

What is the benefit of taking these strange-sounding reexpressions? On the froot scale, the percentages are approximately symmetric, and symmetric data has a clear "average." On the froot scale, a typical Top Ten percentage is 5.7. Also this reexpression can help to equalize spreads between groups, and provide a simple comparison. To illustrate, we use the subset function to create a data frame college34 with data from the Tier 3 and Tier 4 colleges. (The "|" symbol is the logical "or" operator; here we wish to include colleges that are either in Tier 3 or Tier 4.) We compute froots of the Top Ten percentages and use parallel boxplots to compare the two Tiers.

```
> college34 = subset(college, Tier==3 | Tier==4)
> froot = sqrt(college34$Top.10) - sqrt(100 - college34$Top.10)
> boxplot(froot ~ Tier, data=college34, horizontal=TRUE,
+    xlab="Froot(Top 10 Pct)", ylab="Tier")
```

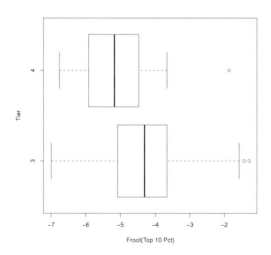

**Fig. 5.11** Parallel boxplots of the froot Top 10 percentages of the Tier 3 and Tier 4 National universities.

We see from the display in Figure 5.11 that the Top 10 percentages, on the froot scale, have similar spreads that we measure by the quartile spread. One can compute the medians of the Top 10 froot percentages to be respectively $-4.3$ and $-5.2$. Therefore, on the froot scale, the Top Ten percentages for the Tier 3 schools tend to be $-4.3 - (-5.2) = 0.9$ higher than the Tier 4 schools.

## Exercises

**5.1 (Exploring percentages of small classes).** The variable Pct.20 in the college dataset contains the percentage of "small classes" (defined as 20 or fewer students) in the National Universities.

a. Construct a dotplot of the small-class percentages using the stripchart function. To see the density of points, it is helpful to use either the method=stack or method=jitter arguments. What is the shape of this data?
b. There is a single school with an unusually large small-class percentage. Use the identify function to find the name of this unusual school.
c. Find the median small-class percentage and draw a vertical line (using the abline function) on the dotplot at the location of the median.

**5.2 (Relationship between the percentages of small classes and large classes).** The variables Pct.20 and Pct.50 in the college dataset contain respectively the percentage of "small classes" (defined as 20 or fewer students) and the percentage of "large classes" (defined as 50 or more students) in the National Universities.

a. Use the plot function to construct a scatterplot of Pct.20 (horizontal) against Pct.50 (vertical).
b. Use the line function to find a resistant line to these data. Add this resistant line to the scatterplot constructed in part a.
c. If 60% of the classes at a particular college have 20 or fewer students, use the fitted line to predict the percentage of classes that have 50 or more students.
d. Construct a graph of the residuals (vertical) against Pct.20 (horizontal) and add a horizontal line at zero (using the abline function).
e. Is there a distinctive pattern to the residuals? (Compare the sizes of the residuals for small Pct.20 and the sizes of the residuals for large Pct.50.)
f. Use the identify function to identify the schools that have residuals that exceed 10 in absolute value. Interpret these large residuals in the context of the problem.

**5.3 (Relationship between acceptance rate and "top-ten" percentage).** The variables Accept.rate and Top.10 in the college dataset contain respectively the acceptance rate and the percentage of incoming students in the top 10 percent of their high school class in the National Universities. One would believe that these two variables are strongly associated, since, for example, "exclusive" colleges with small acceptance rates would be expected to have a large percentage of "top-ten" students.

a. Explore the relationship between `Accept.rate` and `Top.10`. This exploration should include a graph and linear fit that describe the basic pattern in the relationship and a residual graph that shows how schools differ from the basic pattern.
b. Schools are often classified into "elite" and "non-elite" colleges depending on the type of students they admit. Based on your work in part a, is there any evidence from `Accept.rate` and `Top.10` that schools do indeed cluster into "elite" and "non-elite" groups? Explain.

**5.4 (Exploring the pattern of college enrollment in the United States).** The U.S. National Center for Education Statistics lists the total enrollment at Institutions of Higher Education for years 1900-1985 at their website `http://nces.ed.gov`. Define the ordered pair $(x, y)$, where $y$ is the total enrollment in thousands in year $x$. Then we observe the data (1955, 2653), (1956, 2918), (1957, 3324), (1959, 3640), (1961, 4145), (1963, 4780), (1964, 5280), (1965, 5921), (1966, 6390), (1967, 6912), (1968, 7513), (1969, 8005), (1970, 8581).

a. Enter this data into R.
b. Use the `lm` function to fit a line to the pattern of enrollment growth in the period 1955 to 1970. By inspecting a graph of the residuals, decide if a line is a reasonable model of the change in enrollment.
c. Transform the enrollment by a logarithm, and fit a line to the (year, log enrollment) data. Inspect the pattern of residuals and explain why a line is a better fit to the log enrollment data.
d. By interpreting the fit to the log enrollment data, explain how the college enrollment is changing in this time period. How does this growth compare to the growth of the BGSU enrollment in Section 5?

**5.5 (Exploring percentages of full-time faculty).** The variable `Full.time` in the college dataset (see Example 5.3) contains the percentage of faculty who are hired full-time in the group of National Universities.

a. Using the `hist` function, construct a histogram of the full-time percentages and comment on the shape of the distribution.
b. Use the froot and flog transformations to reexpress the full-time percentages. Construct histograms of the collection of froots and the collection of flogs. Is either transformation successful in making the full-time percentages approximately symmetric?
c. For data that is approximately normally distributed, about 68% of the data fall within one standard deviation of the mean. Assuming you have found a transformation in part (b) that makes the full-time percentages

approximately normal, find an interval that contains roughly 68% of the
data on the new scale.

**5.6 (Exploring alumni giving rates).** The variable `Alumni.giving` contains the percentage of alumni from the college who make financial contribu-
tions.

a. Construct a "stacked" dotplot of the alumni giving percentages using the
   `stripchart` function.
b. Identify the names of the three schools with unusually large giving percentages.
c. It can be difficult to summarize these giving percentages since the distribution is right-skewed. One can make the dataset more symmetric by
   applying either a square root transformation or a log transformation.

```
roots = sqrt(college$Alumni.giving)
logs = log(college$Alumni.giving)
```

Apply both square root and log transformations. Which transformation
makes the alumni giving rates approximately symmetric?

**5.7 (Exploring alumni giving rates (continued)).** In this exercise, we
focus on the comparison of the alumni giving percentages between the four
tiers of colleges.

a. Using the `stripchart` function with the stacked option, construct parallel
   dotplots of alumni giving by tier.
b. As one moves from Tier 4 to Tier 1, how does the average giving change?
c. As one moves from Tier 4 to Tier 1, how does the spread of the giving
   rates change?
d. We note from parts (b) and (c), that small giving rates tend to have
   small variation, and large giving rates tend to have large variation. One
   way of removing the dependence of average with spread is to apply a
   power transformation such as a square root or a log. Construct parallel
   stripcharts of the square roots of the giving rates, and parallel boxplots of
   the log giving rates.
e. Looking at the two sets of parallel stripcharts in part (d), were the square
   root rates or the log rates successful in making the spreads approximately
   the same between groups?

# Chapter 6
# Basic Inference Methods

## 6.1 Introduction

*Example 6.1 (Sleeping patterns of college students).*

To illustrate some basic inferential methods, suppose a college instructor is interested in the sleeping patterns of students in a particular mathematics class. He has read that the recommended hours of sleep for a teenager is nine hours each night. That raises several questions:

- Is the median sleeping time for students in this course nine hours?
- If the answer to the first question is no, what proportion of students do get at least nine hours of sleep in a particular night?
- What is a reasonable estimate of the average number of hours these math students get per night?

The instructor decides to collect some data from one representative class to answer these questions. Each student is asked what time he or she got to bed the previous night, and what time he or she woke up the next morning. Based on the answers to these questions, the instructor computes the number of hours of sleep for each of 24 students in his class. The sleeping times are placed in the vector `sleep`.

```
> sleep = c(7.75, 8.5, 8, 6, 8, 6.33, 8.17, 7.75,
+ 7, 6.5, 8.75, 8, 7.5, 3, 6.25, 8.5, 9, 6.5,
+ 9, 9.5, 9, 8, 8, 9.5)
```

In the next sections, this data is analyzed to investigate the sleeping patterns of students.

## 6.2 Learning About a Proportion

### 6.2.1 Testing and estimation problems

Let $M$ denote the median hours of sleep for the population of students who take this math course. We are interested in testing the hypothesis $H$ that $M = 9$ hours. This testing problem can be restated as a test of a population proportion. Let $p$ denote the proportion of students who get at least nine hours of sleep on a particular night. If the population median is $M = 9$ hours, then the proportion $p = 0.5$. So we are interested in testing the hypothesis

$$H : p = 0.5.$$

In the event that $H$ is rejected, one typically is interested in learning about the location of the proportion, and one constructs an interval estimate that contains $p$ with a given confidence.

### 6.2.2 Creating group variables by the ifelse function

The relevant data for this hypothesis test is the sample size and the number of students in the sample who get at least nine hours of sleep. Using the ifelse function, we define a new variable nine.hours that records for each observation if the student got at least nine hours of sleep ("yes") or didn't ("no"). Then we tabulate this "yes, no" data by the table function.

```
> nine.hours = ifelse(sleep >= 9, "yes", "no")
> table(nine.hours)
nine.hours
 no yes
 19   5
```

Only five out of 24 students indicated that they had at least nine hours of sleep. If $H$ is true, the number of yes's has a binomial($n = 24, p = 0.5$) distribution with mean $np$ and variance $np(1-p)$. In addition, if $n$ is large, this variable is approximately normally distributed.

### 6.2.3 Large-sample test and estimation methods

The traditional test for a proportion is based on the assumption that, when the population proportion is $p = 0.5$, the number of yes's $y$ in a sample of $n$ is approximately normally distributed with mean $n/2$ and standard deviation $\sqrt{n/4}$. The $Z$ statistic

$$Z = \frac{y - np}{\sqrt{np(1-p)}},$$

is approximately standard normal. One computes the statistic $z_{obs}$ from the sample and one decides whether to accept or reject $H$ by computing the lower tail probability $P(Z \leq z_{obs})$. If the alternative hypothesis is that $p < 0.5$, the $p$-value is equal to the lower tail probability; if the alternative is two-sided where $p \neq 0.5$, the $p$-value is double the lower-tail probability.

This traditional $Z$ test is implemented using the **prop.test** function. We first define **y** to be the number of yes's and **n** to be the sample size. In the **prop.test** function, we indicate by the p=0.5 argument that we are testing the hypothesis that the proportion is equal to 0.5, and **correct=FALSE** indicates that no continuity correction is used in the calculation of the $Z$ statistic. The summary of this test is displayed by printing the variable **Test**.

```
> y = 5; n = 24
> Test = prop.test(y, n, p=0.5, alternative="two.sided",
+    correct=FALSE)
> Test
        1-sample proportions test without continuity correction

data:  y out of n, null probability 0.5
X-squared = 8.1667, df = 1, p-value = 0.004267
alternative hypothesis: true p is not equal to 0.5
95 percent confidence interval:
 0.09244825 0.40470453
sample estimates:
        p
0.2083333
```

The variable **Test** contains all of the calculations of the test and we request components of **Test** to obtain specific quantities of interest. A vector of the names of components is obtained using the **names** function.

```
> names(Test)
[1] "statistic"   "parameter"   "p.value"     "estimate"    "null.value"
[6] "conf.int"    "alternative" "method"      "data.name"
```

The estimate of the proportion $p$ is the sample proportion of students $y/n$ obtained by asking for the component **estimate**.

```
> Test$estimate
        p
0.2083333
```

The component **statistic** gives the value of the chi-square statistic $z_{obs}^2$ (the square of the observed $Z$ statistic) and **p.value** gives the associated $p$-value. It is a two-sided $p$-value since we indicated the alternative was two-sided.

```
> Test$statistic
X-squared
 8.166667
> Test$p.value
[1] 0.004266725
```

Since the $p$-value is close to zero, we have strong evidence to say that the proportion of "nine hours or greater" sleepers among the students is not 0.5.

In the case where the hypothesis $p = 0.5$ is rejected, a next step is to estimate the proportion by a confidence interval. The component `conf.int` displays a 95% confidence interval. This particular interval, the Wilson score interval, is found by inverting the score test for the proportion.

```
> Test$conf.int
[1] 0.09244825 0.40470453
attr(,"conf.level")
[1] 0.95
```

We are 95% confident that the interval (0.092, 0.405) contains the proportion of heavy sleepers.

### 6.2.4 Small sample methods

One problem with the traditional inference method is that the $Z$ statistic is assumed to be normally distributed, and the accuracy of this normal approximation can be poor for small samples. So there are several alternative inferential methods for a proportion that have better sampling properties when the sample size $n$ is small.

One "small-sample" method is to adjust the $Z$ for the fact that $y$ is a discrete variable. In our example, the "continuity-adjusted" $Z$ statistic for testing the hypothesis $H : p = 0.5$ is based on the statistic

$$Z_{adj} = \frac{y + 0.5 - np}{\sqrt{np(1-p)}}.$$

This test is implemented using the `prop.test` function with the `correct=TRUE` argument.

```
> y = 5; n = 24
> Test.adj = prop.test(y, n, p=0.5, alternative="two.sided",
+    correct=TRUE)
> c(Test.adj$stat, p.value=Test.adj$p.value)
 X-squared      p.value
7.04166667 0.00796349
```

Note that we obtain slightly different values of the chi-square test statistic $Z^2$ and associated $p$-value. The result of 5 successes in 24 trials is slightly less significant using this test.

A second alternative testing method is based on the underlying exact binomial distribution. Under the hypothesis $H : p = 0.5$, the number of successes $y$ has a binomial distribution with parameters $n = 24$ and $p = 0.5$ and the exact (two-sided) $p$-value is given by

$$2 \times P(y \le 5 | p = 0.5).$$

This procedure is implemented using the function `binom.test`. The inputs are the number of successes, the sample size, and the value of the proportion under the null hypothesis.

```
> Test.exact = binom.test(y, n, p=0.5)
> c(Test.exact$stat, p.value=Test.exact$p.value)
number of successes            p.value
     5.000000000           0.006610751
```

One can check the computation of the $p$-value using the binomial cumulative distribution function `pbinom`. The probability that $y$ is at most 5 is given by `pbinom(5, size=24, prob=0.5)` and so the exact $p$-value is given by

```
> 2 * pbinom(5, size=24, prob=0.5)
[1] 0.006610751
```

which agrees with the output of `binom.test`. One can also obtain the "exact" 95% Clopper-Pearson confidence interval by displaying the component `conf.int`.

```
> Test.exact$conf.int
[1] 0.07131862 0.42151284
attr(,"conf.level")
[1] 0.95
```

This particular confidence interval is guaranteed to have 95% coverage, but it can be a bit longer than other computed "95% intervals."

A third popular "small-sample" confidence interval was developed by Agresti and Coull [2]. A 95% interval is constructed by simply adding two successes and two failures to the dataset and then using the simple formula

$$\tilde{p} - 1.96se, \tilde{p} + 1.96se,$$

where $\tilde{p} = (y+2)/(n+4)$ and $se$ is the usual standard error based on the modified data $se = \sqrt{\tilde{p}(1-\tilde{p})/(n+4)}$. There is no R function in the base package that computes the Agresti-Coull interval, but it is straightforward to write a function to compute this interval. In the user-defined function `agresti.interval` below, the inputs are the number of successes y, the sample size n, and the confidence level.

```
agresti.interval = function(y, n, conf=0.95){
  n1 = n + 4
  y1 = y + 2
  phat = y1 / n1
  me = qnorm(1 - (1 - conf) / 2) * sqrt(phat * (1 - phat) / n1)
  c(phat - me, phat + me)
}
```

After this function has been read into R, one computes the interval for our data by typing

```
> agresti.interval(y, n)
[1] 0.0896128 0.4103872
```

(The function `add4ci` in the `PropCIs` package will also compute the Agresti-Coull interval.)

We have illustrated three methods for constructing a 95% interval estimate for a proportion. In the following code, we create a data frame in R that gives the method, the function for implementing the method, and the lower and upper bounds for the interval.

```
> cnames = c("Wilson Score Interval", "Clopper-Pearson",
+ "Agresti-Coull")
> cfunctions = c("prop.test", "binom.test", "agresti.interval")
> intervals = rbind(Test$conf.int, Test.exact$conf.int,
+   agresti.interval(y, n))
> data.frame(Name=cnames, Function=cfunctions,
+   LO=intervals[ , 1], HI=intervals[ , 2])
                   Name         Function         LO        HI
1 Wilson Score Interval        prop.test 0.09244825 0.4047045
2        Clopper-Pearson       binom.test 0.07131862 0.4215128
3          Agresti-Coull agresti.interval 0.08961280 0.4103872
```

In terms of interval length, the shortest interval is the Wilson score interval, followed in order by the Agresti-Coull interval and the Clopper-Pearson interval. Although it is desirable to have short intervals, one wants the interval to have the stated 95% coverage probability. In Chapter 13, we will explore the coverage probability of these three procedures.

# 6.3 Learning About a Mean

## 6.3.1 Introduction

In the previous section, we focused on the proportion of students who got at least nine hours of sleep and learned that this proportion was quite small. Next it is reasonable to return to the original collection of observed sleeping times and learn about the mean $\mu$ of the population of sleeping times for all college students.

## 6.3.2 One-sample t statistic methods

The R function `t.test` performs the calculations for the traditional inference procedures for a population mean. If the observations represent a random sample from a normal population with unknown mean $\mu$, the statistic

$$T = \frac{\sqrt{n}(\bar{y} - \mu)}{s}$$

has a t distribution with $n - 1$ degrees of freedom, where $\bar{y}$, $s$, and $n$ are respectively the sample mean, sample standard deviation, and sample size.

To illustrate this function, suppose we wish to test the hypothesis that the mean sleeping time is 8 hours and also construct a 90% interval estimate for the population mean. Before we use **t.test**, we should check if it is reasonable to assume the sleeping times are normally distributed. A histogram of the times is constructed by the **hist** function. The **qqnorm** function produces a normal probability plot of the times and the **qqline** function overlays a line on the plot passing through the first and third quartiles. The resulting graphs are displayed in Figure 6.1.

```
> hist(sleep)
> qqnorm(sleep)
> qqline(sleep)
```

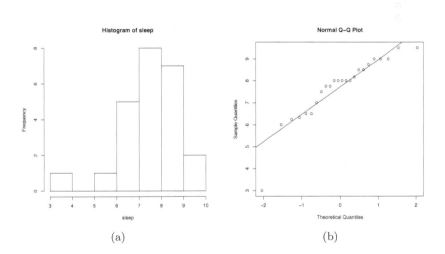

(a)                                         (b)

**Fig. 6.1** Histogram and normal probability plot of the sleeping times of college students.

Looking at the graphs in Figure 6.1, it is clear that there is one unusually small sleeping time (about 3 hours) that is not consistent with the normal distribution assumption. There are several ways one could handle this problem. If there was some mistake in the recording of this particular time, one could remove the outlier and apply the t methods to the modified data. Alternatively, one could apply a different inferential procedure that rests on a more general assumption about the distribution of the population. Here we will illustrate both approaches.

We identify the position of the outlier by constructing an index plot of the sleeping times by the `plot` function. From the display in Figure 6.2, it is clear that the 14th observation is the outlier, and in further examination, we find that this observation was incorrectly coded. A new data vector `sleep.new` is defined that is the original dataset with the 14th observation deleted.

```
> plot(sleep)
> sleep.new = sleep[-14]
```

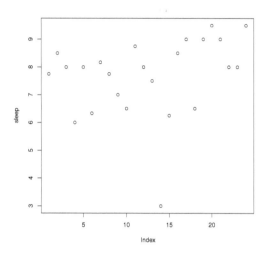

**Fig. 6.2** Index plot of the sleeping times. This helps identify the outlier as the 14th observation.

With the outlier removed, the assumption that the data comes from a normal population seems reasonable, and we can apply the procedures based on the t distribution. The general form of the `t.test` function is given by

```
t.test(x, y=NULL,
       alternative=c("two.sided", "less", "greater"),
       mu=0, paired=FALSE, var.equal=FALSE,
       conf.level=0.95, ...)
```

For single sample inference, the data is contained in the vector x. We indicate by the `mu` argument what value of the population mean is to be tested and the `alternative` argument indicates if the alternative hypothesis is two-sided, less than the population mean value, or greater than the mean value. Also the `conf.level` argument indicates the confidence interval for our interval estimate. In our example, we are interested in testing the hypothesis $\mu = 8$ hours, the alternative is two-sided (the default value), and we wish to construct a 90% interval. The form of the `t.test` function is as follows.

```
> t.test(sleep.new, mu=8, conf.level=0.90)
          One Sample t-test

data:  sleep.new
t = -0.4986, df = 22, p-value = 0.623
alternative hypothesis: true mean is not equal to 8
90 percent confidence interval:
 7.516975 8.265633
sample estimates:
mean of x
 7.891304
```

The output gives the value of the t-test statistic and the two-sided $p$-value for testing the hypothesis that $\mu = 8$. Since the $p$-value is large, there is insufficient evidence from the data to conclude the mean sleeping time of students is not equal to 8 hours. From a frequentist perspective, we are 90% confident that the interval (7.52, 8.27) contains the mean $\mu$.


## 6.3.3 Nonparametric methods

If we wish to use the complete sleeping dataset, it would be inappropriate to use the t procedures due to the single outlier. But there are alternative inferential procedures we can use based on less restrictive assumptions about the population of sleeping times. The Wilcoxon signed rank procedure makes the general assumption that the population is symmetric about a median $M$. In this setting, if one wishes to test the hypothesis that the median sleeping time $M = 8$ hours, then the test statistic is obtained by ranking the absolute values of the differences of each sample value from 8, and then computing the sum of the ranks of the differences that are positive. If the actual median is different from 8, then the sum of ranks corresponding to the positive differences will be unusually small or large, so one rejects the hypothesis in the tail region of the null distribution of the Wilcoxon statistic.

The Wilcoxon signed rank method is implemented as `wilcox.test` with the following general syntax.

```
wilcox.test(x, y=NULL,
            alternative=c("two.sided", "less", "greater"),
            mu=0, paired=FALSE, exact=NULL, correct=TRUE,
            conf.int=FALSE, conf.level=0.95, ...)
```

The arguments are similar to those in `t.test`. The vector `x` is the sample of observations, the constant `mu` is the value to be tested, and the **alternative** argument indicates the direction of the alternative hypothesis. By indicating `conf.int = TRUE`, one can have the function compute the Wilcoxon signed-rank interval estimate for the median and `conf.level` indicates the desired probability of coverage.

We can test the hypothesis $M = 8$ with a two-sided alternative and obtain a 90% interval estimate for the median of sleeping times (with the original dataset) using the command

```
> W = wilcox.test(sleep, mu=8, conf.int=TRUE, conf.level=0.90)
Warning messages:
1: In wilcox.test.default(sleep, mu = 8, conf.int = TRUE, conf.level = 0.9) :
  cannot compute exact $p$-value with ties

> W
          Wilcoxon signed rank test with continuity correction

data:  sleep
V = 73.5, p-value = 0.3969
alternative hypothesis: true location is not equal to 8
90 percent confidence interval:
 7.124979 8.374997
sample estimates:
(pseudo)median
      7.749961
```

We see that a warning message is produced when we execute this function. For small samples (such as this one), the `wilcox.test` will compute exact $p$-values and interval estimates, but these exact methods cannot be used when there are ties in the dataset. When there are ties, as in this sample of sleeping times, the function will give $p$-values and interval estimates based on a normal approximation to the signed-rank statistic.

Using the **names** function, one can see the names of the components of the object producted by the `wilcox.test` function.

```
> names(W)
[1] "statistic"   "parameter"   "p.value"    "null.value"  "alternative"
[6] "method"      "data.name"   "conf.int"   "estimate"
```

The **statistic** component gives the value of the Wilcoxon test statistic, the **p.value** component gives the (two-sided, in this case) $p$-value and the **conf.int** component contains the interval estimate.

```
> W$statistic
   V
73.5
> W$p.value
[1] 0.3968656
> W$conf.int
[1] 7.124979 8.374997
attr(,"conf.level")
[1] 0.9
```

Comparing with the output from `t.test`, both the t and Wilcoxon methods indicate insufficient evidence that the "average" sleeping time from the population is not 8 hours. The 90% Wilcoxon interval estimate for the population median is wider than the t interval estimate for the population mean.

# 6.4 Two Sample Inference

## 6.4.1 Introduction

*Example 6.2 (The twins dataset (continued)).*

A basic inferential problem is to compare the locations of two continuous-valued populations. To illustrate different "two-sample" methods, we consider data from Ashenfelter and Krueger [3] who were interested in relating people's education and income. It can be difficult to learn about the effect of education on income since there are many variables associated with income, such as a person's natural ability, his family background, and his innate intelligence. To control for possible confounding variables, the authors collected information on education, income, and background from a group of twins. Since twins have similar family backgrounds, they provide a useful control for confounding variables in this problem. (This particular dataset was previously used in Chapter 3 to illustrate statistical methods for categorical data.)

The datafile `twins.txt` contains information about 183 pairs of twins for sixteen variables. We read the data into R by the `read.table` function and store the data frame in the variable `twins`.

```
> twins = read.table("twins.txt", header=TRUE)
```

Each pair of twins was randomly assigned the labels "twin 1" and "twin 2." The variable `HRWAGEH` gives the hourly wage for twin 2. If one graphs the wages, one finds that they are strongly right-skewed and one can remove the skewness by a log transformation; the variable `log.wages` contains the log wages.

```
> log.wages = log(twins$HRWAGEH)
```

## 6.4.2 Two sample t-test

The variable `EDUCH` contains the self-reported education of twin 2 in years. Suppose one is interested in comparing the log wages of the "high school" twins with 12 or fewer years of education with the log wages of the "college" twins with more than 12 years of education. We define a new categorical variable `college` using the `ifelse` function that is "yes" or "no" depending on the years of education.

```
> college = ifelse(twins$EDUCH > 12, "yes", "no")
```

A first step in comparing the log wages of the two groups is to construct a suitable graph and Figure 6.3 displays parallel boxplots of the log wages using the `boxplot` function. Both groups of log wages look approximately symmetric with similar spreads and the median log wage of the college twins appears approximately 0.5 larger than the median log wage of the high school twins.

```
> boxplot(log.wages ~ college, horizontal=TRUE,
+   names=c("High School", "Some College"), xlab="log Wage")
```

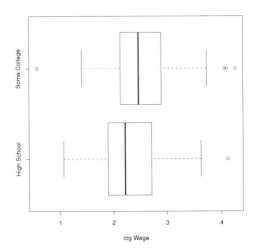

**Fig. 6.3** Parallel boxplots of the log wages for the high school and college twins.

Let $\mu_H$ and $\mu_C$ denote respectively the mean log wage of the population of high school twins and college twins. The standard t-test of the hypothesis $H$ that $\mu_H = \mu_C$ is implemented by the function `t.test`. The argument has the form `log.wages ~ college` where the `log.wages` is the continuous response variable and `college` categorizes the response into two groups.

```
> t.test(log.wages ~ college)
        Welch Two Sample t-test

data:  hs.log.wages and college.log.wages
t = -2.4545, df = 131.24, p-value = 0.01542
alternative hypothesis: true difference in means is not equal to 0
95 percent confidence interval:
 -0.42999633 -0.04620214
sample estimates:
 mean in group no mean in group yes
         2.282119          2.520218
```

From the output, we see that:

• The value of the t-test statistic is $-2.4545$.

- Using the Welch procedure for unequal population variances, the t statistic is approximately distributed as t with 131.24 degrees of freedom under the assumption of equal population means.
- The two-sided $p$-value is 0.01542, so there is significant evidence that the means are different.
- The 95% confidence interval for the difference in means $\mu_H - \mu_C$ is $(-0.430, -0.046)$.

The traditional t-test for the difference of means assumes the population variances are equal. One can implement this traditional test using the `var.equal=TRUE` argument.

```
> t.test(log.wages ~ college, var.equal=TRUE)$p.value
[1] 0.01907047
```

In this example, the Welch test and the traditional t-test give approximately the same $p$-value.

### 6.4.3 Two sample Mann-Whitney-Wilcoxon test

A nonparametric alternative to the two-sample test is the Mann-Whitney-Wilcoxon test. One tests the general hypothesis that two independent samples come from the same continuous population. If one denotes the two samples as $x$ and $y$, the test statistic is equal to

$$W = \text{the number of pairs } (x_i, y_j) \text{ where } x_i > y_j.$$

This method is implemented using the `wilcox.test` function with the same argument format (response by grouping variable) as the `t.test` function.

```
> wilcox.test(log.wages ~ college, conf.int=TRUE)
        Wilcoxon rank sum test with continuity correction

data:  hs.log.wages and college.log.wages
W = 2264, p-value = 0.01093
alternative hypothesis: true location shift is not equal to 0
95 percent confidence interval:
 -0.44266384 -0.06455011
sample estimates:
difference in location
             -0.2575775
```

The value of the $W$ statistic for our dataset is 2264. The two-sided probability of getting a value as extreme as 2264 if the two samples come from the same continuous population is 0.01093. This $p$-value is very close to the value obtained using the two-sample t-test.

By specifying the argument option `conf.int = TRUE`, `wilcox.test` will also display a 95% confidence interval for the difference of location parameters of the two populations. Comparing the output of `wilcox.test` with the output of `t.test`, we see this confidence interval is similar to the interval for the difference of population means.

### 6.4.4 Permutation test

Another testing procedure for comparing two independent samples is a permutation test. As before, we define a variable `log.wages` that contains the log wages for the first twin and a vector `college` that indicates if the first twin had some college education or not.

```
> log.wages = log(twins$HRWAGEH)
> college = ifelse(twins$EDUCH > 12, "yes", "no")
```

Using the `table` function, we see that there are 112 college and 71 no-college twins in our sample.

```
> table(college)
college
 no yes
 71 112
```

Consider the hypothesis that college education has no impact on the wages of the twins. Under this hypothesis, the labeling of the twins into the college and no-college categories is not helpful in understanding the variability of the log wages. In this case, the distribution of any test statistic, say the t-test, will be unchanged if we arbitrarily change the labels of the twins in the college/no-college classification. One obtains an empirical distribution of the test statistic under the null hypothesis by

- Randomly allocating the 71 college and 112 no-college labels among the 183 twins.
- Computing the value of the test statistic for the randomly permuted data.
- Repeating this process a large number of iterations.

The collection of test statistics provides an estimate of the sampling distribution of the statistic when the null hypothesis is true. (Figure 6.4 is a picture of this sampling distribution.) The observed value of the test statistic (for our original sample) is then compared to the distribution of the permutation replicates. To make this comparison, one computes the probability that the test statistic under the randomization distribution is at least as extreme as the observed statistic. If this $p$-value is sufficiently small, this gives evidence against the assumption that the labeling of the twins into the two groups is not informative.

This testing procedure is straightforward to program by writing a short function. The function `resample` is written which randomly permutes the college labels (using the `sample` function) and returns the value of the t-test statistic from the `t.test` function.

```
> resample = function()
+    t.test(log.wages ~ sample(college))$statistic
```

One repeats `resample` using the `replicate` function. There are two arguments to `replicate`, the number of iterations and the name of the function to be replicated. The values of the t statistic from the 1000 iterations is stored in the vector `many.T`.

```
> many.T = replicate(1000, resample())
```

The value of the t statistic from the observed data is obtained by the `t.test` function using the labels from the vector `college`.

```
> T.obs = t.test(log.wages ~ college)$statistic
> T.obs
       t
-2.454488
```

To see if the observed test statistic of $-2.45$ is extreme, we use the `hist` function to construct a histogram of the t statistics from the randomized distribution in Figure 6.4. We use the `abline` function to add a vertical line at the observed t statistic.

```
> hist(many.T)
> abline(v=T.obs)
```

The (two-sided) $p$-value is twice the probability of obtaining a t statistic smaller than `T.obs`.

```
> 2 * mean(many.T < T.obs)
[1] 0.024
```

This computed $p$-value from the permutation test is similar to the values computed using the t-test and Wilcoxon procedures.

## 6.5 Paired Sample Inference Using a t Statistic

In the above analysis, we were comparing the wages of two groups of people with different educational levels. It is difficult to get precise estimates of the effect of education on wage since there are many other confounding variables such as family background, natural ability, and innate intelligence that may also explain the differences between the two groups. This particular study took a sample of twins. By considering the wages of twins who differ in educational level but are similar with respect to other important variables such

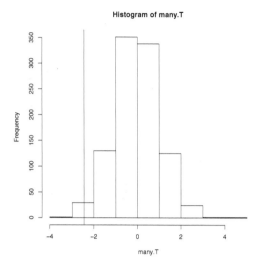

**Fig. 6.4** Simulated randomization distribution of the two sample t statistic under the null hypothesis in the permutation test. The vertical line gives the location of the observed t statistic.

as intelligence and family background, one might more accurately estimate the effect of education. Here we illustrate methods of comparing two means from paired data.

In the dataframe **twins**, the variables **EDUCL** and **EDUCH** give the educational levels for twin 1 and twin 2. We first create a new data frame consisting only of the twins with different educational levels. This new data frame **twins.diff** is created using the **subset** function.

```
> twins.diff = subset(twins, EDUCL != EDUCH)
```

(The != symbol means "not equal.") Since there are twins with missing values of educational level, we use the **complete.cases** function to remove any rows from the dataframe where any of the variables contain a **NA** code.

```
> twins.diff = twins.diff[complete.cases(twins.diff), ]
```

For these twins with different educational levels, we let **log.wages.low** be the log hourly wage for the twin with the lower educational level and **log.wages.high** the log hourly wage for the twin with the higher educational level. We compute these new variables using of the **ifelse** function. For example, if the condition EDUCL < EDUCH is true, then **log.wages.low** will be equal to **log(HRWAGEL)**; otherwise **log.wages.low** will be equal to **log(HRWAGEH)**. A similar conditional expression is used to compute the variable **log.wages.high**.

```
> log.wages.low = with(twins.diff,
+    ifelse(EDUCL < EDUCH, log(HRWAGEL), log(HRWAGEH)))
```

```
> log.wages.high = with(twins.diff,
+   ifelse(EDUCL < EDUCH, log(HRWAGEH), log(HRWAGEL)))
```

When we are done with this data cleaning, we have data on log wages for 75 pairs of twins. We combine the twins data by the cbind function and display the log wages for the first six pairs of twins using of the head function.

```
> head(cbind(log.wages.low, log.wages.high))
  log.wages.low log.wages.high
1      2.169054       2.890372
2      3.555348       2.032088
3      2.484907       2.708050
4      2.847812       2.796061
5      2.748872       3.218876
6      2.079442       2.708050
```

Let $\mu_L$ and $\mu_H$ denote respectively the mean log wages for the twins with the lower and higher educational levels. Due to the paired design, one can perform a test for the difference in means $d = \mu_L - \mu_H$ by working with the single sample of paired differences log.wages.low - log.wages.high. We use the hist function to construct a histogram of the paired differences in Figure 6.5. Since the paired differences in log wages look approximately

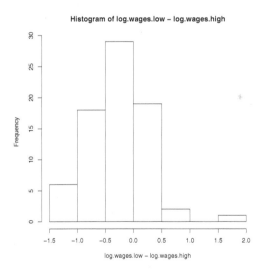

**Fig. 6.5** Histogram of the paired differences to see the effect of education on log wages.

normal, it is reasonable to apply a t-test on the differences by the t.test function. To construct a test based on the paired differences, the argument option paired = TRUE is used.

```
> t.test(log.wages.low, log.wages.high, paired=TRUE)
        Paired t-test

data:  log.wages.low and log.wages.high
t = -4.5516, df = 74, p-value = 2.047e-05
alternative hypothesis: true difference in means is not equal to 0
95 percent confidence interval:
 -0.3930587 -0.1537032
sample estimates:
mean of the differences
             -0.2733810
```

Here the observed difference in means is clearly statistically significant and a
95% confidence interval for the difference is $(-0.393, -0.154)$. In an exercise,
one will be asked to rerun the t.test function on this dataset without using
the paired = TRUE option and comment on the difference in the confidence
intervals for the difference in means using the two options.

# Exercises

**6.1 (Gender of marathoners).** In 2000, the proportion of females who
competed in marathons in the United States was 0.375. One wonders if the
proportion of female marathoners has changed in the ten-year period from
2000 to 2010. One collects the genders of 276 people who competed in the
2010 New York City Marathon – in this sample, 120 were women.

a. If $p$ denotes the proportion of 2010 marathoners who are female, use the
   prop.test function to test the hypothesis that $p = 0.375$. Store the calculations of the test in the variable Test.
b. From the components of Test, construct a 95% interval estimate for $p$.
c. Using the function binom.test, construct an exact-test of the hypothesis.
   Compare this test with the large-sample test used in part (a).

**6.2 (Ages of marathoners).** The datafile "nyc.marathon.txt" contains the
gender, age, and completion time (in minutes) for 276 people who completed
the 2010 New York City Marathon. It was reported that the mean ages of
men and women marathoners in 2005 were respectively 40.5 and 36.1.

a. Create a new dataframe "women.marathon" that contains the ages and
   completion times for the women marathoners.
b. Use the t.test function to construct a test of the hypothesis that the
   mean age of women marathoners is equal to 36.1.
c. As an alternative method, use the wilcox.test function to test the hypothesis that the median age of women marathoners is equal to 36.1. Compare this test with the t-test used in part (b).
d. Construct a 90% interval estimate for the mean age of women marathoners.

**6.3 (Ages of marathoners, continued).** From the information in the 2005 report, one may believe that men marathoners tend to be older than women marathons.

a. Use the `t.test` function to construct a test of the hypothesis that the mean ages of women and men marathoners are equal against the alternative hypothesis that the mean age of men is larger.
b. Construct a 90% interval estimate for the difference in mean ages of men and women marathoners.
c. Use the alternative Mann-Whitney-Wilcoxon test (function `wilcox.test`) to test the hypothesis that the ages of the men and ages of the women come from populations with the same location parameter against the alternative that the population of ages of the men have a larger location parameter. Compare the result of this test with the t-test performed in part (a).

**6.4 (Measuring the length of a string).** An experiment was performed in an introductory statistics class to illustrate the concept of measurement bias. The instructor held up a string in front of the class and each student guessed at the string's length. The following are the measurements from the 24 students (in inches).

```
22 18 27 23 24 15 26 22 24 25 24 18
18 26 20 24 27 16 30 22 17 18 22 26
```

a. Use the `scan` function to enter these measurements into R.
b. The true length of the string was 26 inches. Assuming that this sample of measurements represents a random sample from a population of student measurements, use the `t.test` function to test the hypothesis that the mean measurement $\mu$ is different from 26 inches.
c. Use the `t.test` function to find a 90% confidence interval for the population mean $\mu$.
d. The t-test procedure assumes the sample is from a population that is normally distributed. Construct a normal probability plot of the measurements and decide if the assumption of normality is reasonable.

**6.5 (Comparing snowfall of Buffalo and Cleveland).** The datafile "buffalo.cleveland.snowfall.txt" contains the total snowfall in inches for the cities Buffalo and Cleveland for the seasons 1968-69 through 2008-09.

a. Compute the differences between the Buffalo snowfall and the Cleveland snowfall for all seasons.
b. Using the `t.test` function with the difference data, test the hypothesis that Buffalo and Cleveland get, on average, the same total snowfall in a season.
c. Use the `t.test` function to construct a 95% confidence interval of the mean difference in seasonal snowfall.

**6.6 (Comparing Etruscan and modern Italian skulls).** Researchers were interested if ancient Etruscans were native to Italy. The dataset "Etruscan-Italian.txt" contains the skull measurements from a group of Etruscans and modern Italians. There are two relevant variables in the dataset: x is the skull measurement and group is the type of skull.

a. Assuming that the data represent independent samples from normal distributions, use the t.test function to test the hypothesis that the mean Etruscan skull measurement $\mu_E$ is equal to the mean Italian skull measurement $\mu_I$.

b. Use the t.test function to construct a 95% interval estimate for the difference in means $\mu_E - \mu_I$.

c. Use the two-sample Wilcoxon procedure implemented in the function wilcox.test to find an alternative 95% interval estimate for the difference $\mu_E - \mu_I$.

**6.7 (President's heights).** In Example 1.2, the height of the election winner and loser were collected for the U.S. Presidential elections of 1948 through 2008. Suppose you are interested in testing the hypothesis that the mean height of the election winner is equal to the mean height of the election loser. Assuming that this data represent paired data from a hypothetical population of elections, use the t.test function to test this hypothesis. Interpret the results of this test.

# Chapter 7
# Regression

## 7.1 Introduction

Regression is a general statistical method to fit a straight line or other model
to data. The objective is to find a model for predicting the dependent variable
(*response*) given one or more independent (*predictor*) variables.

The simplest example is a *simple linear regression model* of $Y$ on $X$, defined
by

$$Y = \beta_0 + \beta_1 X + \varepsilon, \tag{7.1}$$

where $\varepsilon$ is a random error term. The word "simple" means that there is one
predictor variable in the model. The linear model (7.1) describes a straight
line relation between the response variable $Y$ and predictor $X$.

In least squares regression the unknown parameters $\beta_0$ and $\beta_1$ are estimated by minimizing the sum of the squared deviations between the observed
response $Y$ and the value $\hat{Y}$ predicted by the model. If these estimates are
$b_0$ (intercept) and $b_1$ (slope), the estimated regression line is

$$\hat{Y} = b_0 + b_1 X.$$

For a set of data $(x_i, y_i)$, $i = 1, \ldots, n$, the errors in this estimate are $y_i - \hat{y}_i$,
$i = 1, \ldots, n$. Least squares regression obtains the estimated intercept $b_0$ and
slope $b_1$ that minimizes the sum of squared errors: $\sum_{i=1}^{n}(y_i - \hat{y}_i)^2$.

A *multiple linear regression model* has more than one predictor variable
(multiple predictors). Linear models can describe many relations other than
straight lines or planes. Any model that is linear in the parameters is considered a linear model. Thus a quadratic relation $y = \beta_0 + \beta_1 x + \beta_2 x^2$ corresponds
to a linear model with two predictors, $X_1 = X$ and $X_2 = X^2$. The exponential
relation $y = \beta_0 e^{\beta_1 x}$ is not linear, but the relation can be expressed by taking the natural logarithm of both sides. The corresponding linear equation is
$\ln y = \ln \beta_0 + \beta_1 x$.

## 7.2 Simple Linear Regression

### 7.2.1 Fitting the model

*Example 7.1 (cars).* Consider a set of paired observations of speed and stopping distance of cars. Is there a linear relation between stopping distance and speed of a car?

The data set `cars` is one of the data sets installed with R. We `attach` the data set `cars`, and from the help page for `cars` (displayed using `?cars`), we learn that there are 50 observations of `speed` (mph) and `dist` (stopping distance in feet), and that this data was recorded in 1920. Our first step in the analysis is to construct a scatterplot of `dist` vs `speed`, using the `plot` function.

```
> attach(cars)     #attach the data
> ?cars            #display the help page for cars data
> plot(cars)       #construct scatterplot
```

The scatterplot displayed in Figure 7.1 reveals that there is a positive association between distance `dist` and `speed` of cars. The relation between distance and speed could be approximated by a line or perhaps by a parabola. We start with the simplest model, the straight line model. The response variable in this example is stopping distance `dist` and the predictor variable speed is `speed`. To fit a straight line model

$$dist = \beta_0 + \beta_1\, speed + \varepsilon,$$

we need estimates of the intercept $\beta_0$ and the slope $\beta_1$ of the line.

**The lm function and the model formula**

The linear model function is `lm`. This function estimates the parameters of a linear model by the least squares method. A linear model is specified in R by a model `formula`.

The R `formula` that specifies a simple linear regression model $dist = \beta_0 + \beta_1 speed + \varepsilon$ is simply

$$dist \sim speed$$

The model formula is the first argument to the `lm` (linear model) function. In this example, the estimated regression model is obtained by

```
> lm(dist ~ speed)
```

The `lm` command above produces the following output.

```
Call:
lm(formula = dist ~ speed)
```

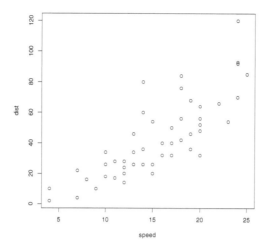

**Fig. 7.1** Scatterplot of stopping distance vs speed in Example 7.1.

```
Coefficients:
(Intercept)        speed
   -17.579         3.932
```

The function `lm` displays only the estimated coefficients, but the object returned by `lm` contains much more information, which we will explore below. As we want to analyze the fit of this model, it is useful to store the result:

```
> L1 = lm(dist ~ speed)
> print(L1)
```

In this case the value of `L1` will be displayed after we type the symbol `L1` or `print(L1)`. The result will be as shown above.

The fitted regression line is: dist $= -17.579 + 3.932$ speed. According to this model, average stopping distance increases by 3.932 feet for each additional mile per hour of speed. The **abline** or **curve** functions can be used to add the fitted line to the scatterplot. See Figure 7.2(a).

```
> plot(cars, main="dist = -17.579 + 3.932 speed", xlim=c(0, 25))
> #line with intercept=-17.579, slope=3.932
> abline(-17.579, 3.932)
> curve(-17.579 + 3.932*x, add=TRUE)   #same thing
```

$\mathbf{R_x}$ **7.1** *A shortcut to add a simple linear regression line to a plot is to supply the result of `lm` as the first argument to `abline`. In the above example, we could have used `abline(lm(dist ~ speed))` or `abline(L1)` to add the fitted line to the plot in Figure 7.2(a).*

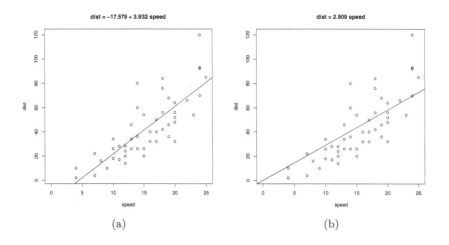

**Fig. 7.2** Regression lines for `speed` vs `dist` in Examples 7.1–7.2. Figure (a) displays the data with the fitted line $\hat{y} = -17.579 + 3.932x$ from Example 7.1. Figure (b) displays the same data with the regression-through-the-origin fit, $\hat{y} = 2.909x$ from Example 7.2.

### 7.2.2 Residuals

The *residuals* are the vertical distances from the observed stopping distance `dist` (the plotting symbol) to the line. The paired observations are $(x_i, y_i) = (\text{speed}_i, \text{dist}_i)$, $i = 1, \ldots, 50$. The residuals are

$$e_i = y_i - \hat{y}_i,$$

where $\hat{y}_i$ denotes the value of stopping distance predicted by the model at speed $x_i$. In this problem, $\hat{y}_i = -17.579 + 3.932x_i$. One can observe in Figure 7.2(a) that the model tends to fit better at slow speeds than at higher speeds. It is somewhat easier to analyze the distribution of the errors using a residual plot. Figure 7.3 is a scatterplot of residuals vs fitted values. One way to generate the residual plot is to use the plot method for the `lm` result. The `which=1` argument specifies the type of plot (residuals vs fitted values). The `add.smooth` argument controls whether a certain type of curve is fitted to the residuals.

```
> plot(L1, which=1, add.smooth=FALSE)
```

The residual plot (Figure 7.3) has three unusually large residuals labeled; observations 23, 35, and 49. One can also observe that the residuals are closer to zero at slow speeds; the variance of the residuals is not constant across all speeds, but increasing with speed. Inference (tests or confidence intervals) about the model is usually based on the assumption that the errors are normally distributed with mean zero and constant variance.

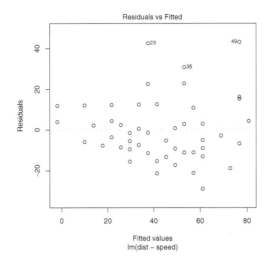

**Fig. 7.3** Scatterplot of residuals vs fitted values for the **cars** data in Example 7.1.

### 7.2.3 Regression through the origin

*Example 7.2 (cars, cont.).* The **cars** data includes speeds as slow as 4 mph, and the estimated intercept should correspond to the expected distance required to stop a car that is not moving; however, our estimated intercept is -17.579 feet. The model with intercept zero

$$Y = \beta_1 X + \varepsilon$$

can be estimated by explicitly including the intercept 0 in the model formula. Then `lm(dist ~ 0 + speed)` sets intercept equal to zero and estimates the slope by the least squares method.

```
> L2 = lm(dist ~ 0 + speed)
> L2
Call:
lm(formula = dist ~ 0 + speed)

Coefficients:
speed
2.909
```

The estimated slope for this model is 2.909. For each additional one mph of speed, the estimated average stopping distance increases 2.909 feet. The fitted line plot for this model is in Figure 7.2(b). It is generated by the following code.

```
> plot(cars, main="dist = 2.909 speed", xlim=c(0,25))
> #line with intercept=0, slope=2.909
> abline(0, 2.909)
```

Again we observe that the fit is better at slow speeds than at faster speeds. A plot of residuals vs fitted values for this model can be generated by

```
> plot(L2, which=1, add.smooth=FALSE)
```

as described above. This plot (not shown) looks very similar to Figure 7.3.

A quadratic model could be considered for this data; see Exercise 7.8.

The cars data can be detached when it is no longer needed, using

```
> detach(cars)
```

## 7.3 Regression Analysis for Data with Two Predictors

In the next example there are two predictor variables. One could fit a simple linear regression model with either of these variables, or fit a multiple linear regression model using both predictors.

### 7.3.1 Preliminary analysis

*Example 7.3 (Volume of Black Cherry Trees).* The data file "cherry.txt" can be obtained from *StatSci* at http://www.statsci.org/data/general/cherry.html and can also be found in Hand et al. [21]. The data were collected from 31 black cherry trees in the Allegheny National Forest, Pennsylvania, in order to find an estimate for the volume of a tree (and therefore the timber yield), given its height and diameter. The data set contains a sample of 31 observations of the variables

| Variable | Description |
|----------|-------------|
| Diam | diameter in inches |
| Height | height in feet |
| Volume | cubic feet |

This data set is also available in R as trees. It is identical to "cherry.txt" except that the diameter variable is named "Girth". We use the R data and rename the diameter as Diam, creating a new data frame called Trees, then attach the data frame.

```
> Trees = trees
> names(Trees)[1] = "Diam"
> attach(Trees)
```

The **pairs** function generates an array of scatterplots for each pair of variables. This type of plot (Figure 7.4) helps visualize the relations between variables.

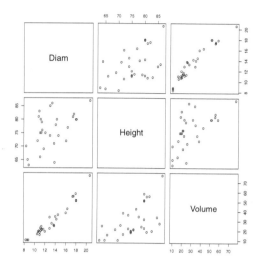

**Fig. 7.4** Pairs plot of cherry tree data in Example 7.3.

In the **pairs** plot (Figure 7.4) the variables **Diam** and **Volume** appear to have a strong linear association, and Height and Volume are also related.

We also print a correlation matrix.

```
> pairs(Trees)
> cor(Trees)
```

```
            Diam     Height    Volume
Diam   1.0000000 0.5192801 0.9671194
Height 0.5192801 1.0000000 0.5982497
Volume 0.9671194 0.5982497 1.0000000
```

The correlation between diameter and volume is 0.97, indicating a strong positive linear relation between **Diam** and **Volume**. The correlation between height and volume is 0.60, which indicates a moderately strong positive linear association between **Height** and **Volume**.

As a first step, let us fit a simple linear regression model with diameter as the predictor:

$$Y = \beta_0 + \beta_1 X_1 + \varepsilon,$$

where $Y$ is the volume, $X_1$ is the diameter, and $\varepsilon$ is random error; call this Model 1. The `lm` function is used to fit the model and we store the result in M1. The intercept term is included in the formula by default.

```
> M1 = lm(Volume ~ Diam)
> print(M1)

Call:
lm(formula = Volume ~ Diam, data = Trees)

Coefficients:
(Intercept)         Diam
    -36.943        5.066
```

The estimated intercept is -36.943 and the estimated slope is 5.066. According to this model, the average volume increases by 5.066 cubic feet for each additional 1 inch in diameter.

The fitted model M1 contains the estimated coefficients. We add the line to the scatterplot of the data using the vector of coefficients `M1$coef`. The scatterplot with fitted line is shown in Figure 7.5

```
> plot(Diam, Volume)      #response vs predictor
> abline(M1$coef)         #add fitted line
```

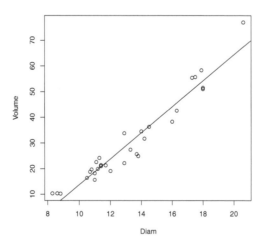

**Fig. 7.5** Scatterplot of cherry tree volume vs tree diameter with fitted regression line in Example 7.3.

To predict volume for new trees, the `predict` method can be used. Store the diameter value for the new tree(s) in a data frame using the name `Diam` of the original model formula specified in the `lm` call. For example, the predicted volume for a new tree with diameter 16 in. is obtained by

```
> new = data.frame(Diam=16)
> predict(M1, new)
```

```
        1
44.11024
```

The predicted volume of the new tree is 44.1 cubic feet.

For inference, one requires some assumptions about the distribution of the error term. We assume that the random errors $\varepsilon$ are independent and identically distributed (iid) as Normal$(0, \sigma^2)$ random variables. Residual plots help us to assess the fit of the model and the assumptions for $\varepsilon$.

One can obtain residual plots using the `plot` method for `lm`; here we are requesting two plots: a plot of residuals vs fits (1) and a QQ plot to check for normality of residuals (2).

```
plot(M1, which=1:2)
```

The user is prompted for each graph with a message at the console:

```
Waiting to confirm page change...
```

The residual plots are shown in Figure 7.6(a) and 7.6(b). In Figure 7.6(a) a curve has been added. This curve is a fitted `lowess` (local polynomial regression) curve, called a *smoother*. The residuals are assumed iid, but there is a pattern evident. The residuals have a "U" shape or bowl shape. This pattern could indicate that there is a variable missing from the model. In the QQ plot 7.6(b), normally distributed residuals should lie approximately along the reference line shown in the plot. The observation with the largest residual corresponds to the tree with the largest volume, observation 31. It also has the largest height and diameter.

## 7.3.2 Multiple regression model

A multiple linear regression model with response variable $Y$ and two predictor variables $X_1$ and $X_2$ is

$$Y = \beta_0 + \beta_1 X_1 + \beta_2 X_2 + \varepsilon,$$

where $\varepsilon$ is a random error term. For inference we assume that the errors are normally distributed and independent with mean zero and common variance $\sigma^2$.

*Example 7.4 (Model for Volume of Cherry Trees, cont.).* Next we consider the two variable model for predicting volume of cherry trees given diameter and height. This is a multiple regression model with two predictors; $X_1$ is the diameter and $X_2$ is the height of the tree. The response variable is $Y$, the volume. Call this Model 2.

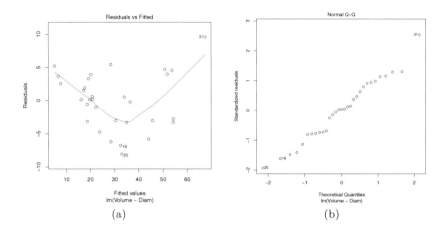

**Fig. 7.6** Residuals vs fits (a) and normal QQ plot of residuals (b) in Example 7.3 for Model 1.

Least squares estimates of the parameters of multiple linear regression models are obtained by lm, using similar syntax as for simple linear regression. The model formula determines which type of model is fit. The model formula we require is

$$\text{Volume} \sim \text{Diam} + \text{Height}$$

and we fit the model, store it as M2, then print the result with the commands

```
> M2 = lm(Volume ~ Diam + Height)
> print(M2)

Call:
lm(formula = Volume ~ Diam + Height)

Coefficients:
(Intercept)        Diam       Height
   -57.9877      4.7082       0.3393
```

The fitted regression model is

$$\hat{Y} = -57.9877 + 4.7082X_1 + 0.3393X_2$$

or Volume $= -57.9877 + 4.7082\,\text{Diam} + 0.3393\,\text{Height} + \text{error}$. According to this model, when height is held constant, average volume of a tree increases by 4.7082 cubic feet for each additional inch in diameter. When diameter is held constant, average volume of a tree increases by 0.3393 cubic feet for each additional inch of height.

The residual plots for Model 2 are obtained by

```
> plot(M2, which=1:2)
```

(see the previous section). These residual plots for Model 2 (not shown) look similar to the corresponding plots for M1 in Figure 7.6(a). The "U" shaped pattern of residuals in the plot of residuals vs fits (similar to Figure 7.6(a)) may indicate that a quadratic term is missing from the model.

*Example 7.5 (Model for Cherry Trees, cont.).* Finally, let us fit a model that also includes the square of diameter as a predictor. Call this Model 3. The model is specified by the formula

```
Volume ~ Diam + I(Diam^2) + Height
```

where I(Diam^2) means to interpret Diam^2 "as is" (the square of Diam) rather than interpret the exponent as a formula operator. We fit the model, storing the result in M3.

```
> M3 = lm(Volume ~ Diam + I(Diam^2) + Height)
> print(M3)

Call:
lm(formula = Volume ~ Diam + I(Diam^2) + Height)

Coefficients:
(Intercept)        Diam     I(Diam^2)        Height
    -9.9204      -2.8851        0.2686        0.3764
```

Then we display the residual plots, which are shown in Figures 7.7(a) and 7.7(b).

```
plot(M3, which=1:2)
```

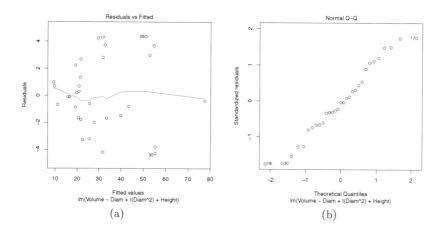

(a)                                             (b)

**Fig. 7.7** Residuals vs fits (a) and normal QQ plot of residuals in Example 7.5 for Model 3.

For Model 3, the plot of residuals vs fits in Figure 7.7(a) does not have the "U" shape that was apparent for Models 1 and 2. The residuals are approximately centered at 0 with constant variance. In the normal QQ plot (Figure 7.7(b)), the residuals are close to the reference line on the plot. These residual plots are consistent with the assumption that errors are iid with a Normal$(0, \sigma^2)$ distribution.

### 7.3.3 The summary and anova methods for lm

The summary of the fitted model contains additional information about the model. In the result of summary we find a table of the coefficients with standard errors, a five number summary of residuals, the coefficient of determination $(R^2)$, and the residual standard error.

*Example 7.6 (Cherry Trees Model 3).* The summary of our multiple regression fit stored in M3 is obtained below.

```
> summary(M3)

Call:
lm(formula = Volume ~ Diam + I(Diam^2) + Height)

Residuals:
    Min      1Q  Median      3Q     Max
-4.2928 -1.6693 -0.1018  1.7851  4.3489

Coefficients:
             Estimate Std. Error t value Pr(>|t|)
(Intercept) -9.92041   10.07911   -0.984 0.333729
Diam        -2.88508    1.30985   -2.203 0.036343 *
I(Diam^2)    0.26862    0.04590    5.852 3.13e-06 ***
Height       0.37639    0.08823    4.266 0.000218 ***
---
Signif. codes:  0 '***' 0.001 '**' 0.01 '*' 0.05 '.' 0.1 ' ' 1

Residual standard error: 2.625 on 27 degrees of freedom
Multiple R-squared: 0.9771,    Adjusted R-squared: 0.9745
F-statistic: 383.2 on 3 and 27 DF,  p-value: < 2.2e-16
```

The adjusted $R^2$ value of 0.9745 indicates that more than 97% of the total variation in Volume about its mean is explained by the linear association with the predictors Diam, Diam$^2$, and Height. The residual standard error is 2.625. This is the estimate of $\sigma$, the standard deviation of the error term $\varepsilon$ in Model 3.

The table of coefficients includes standard errors and $t$ statistics for testing $H_0 : \beta_j = 0$ vs $H_1 : \beta_j \neq 0$. The $p$-values of the test statistics are given under Pr(>|t|). We reject the null hypothesis $H_0 : \beta_j = 0$ if the corresponding $p$-

value is less than the significance level. At significance level 0.05 we conclude that Diam, Diam$^2$, and Height are significant.

The analysis of variance (ANOVA) table for this model is obtained by the anova function.

```
> anova(M3)

Analysis of Variance Table

Response: Volume
          Df Sum Sq Mean Sq F value    Pr(>F)
Diam       1 7581.8  7581.8 1100.511 < 2.2e-16 ***
I(Diam^2)  1  212.9   212.9   30.906 6.807e-06 ***
Height     1  125.4   125.4   18.198 0.0002183 ***
Residuals 27  186.0     6.9
```

From the ANOVA table, one can observe that Diam explains most of the total variability in the response, but the other predictors are also significant in Model 3.

A way to compare the models (Model 1 in Example 7.3, Model 2 in Example 7.4, and Model 3 in Example 7.5) is to list all of the corresponding lm objects as arguments to anova,

```
> anova(M1, M2, M3)
```

which produces the following table:

```
Analysis of Variance Table

Model 1: Volume ~ Diam
Model 2: Volume ~ Diam + Height
Model 3: Volume ~ Diam + I(Diam^2) + Height
  Res.Df    RSS Df Sum of Sq      F    Pr(>F)
1     29 524.30
2     28 421.92  1    102.38 14.861 0.0006487 ***
3     27 186.01  1    235.91 34.243  3.13e-06 ***
---
Signif. codes:  0 '***' 0.001 '**' 0.01 '*' 0.05 '.' 0.1 ' ' 1
```

This table shows that the residual sum of squares decreased by 102.38 from 524.30 when Height was added to the model, and decreased another 235.91 from 421.92 when the square of diameter was added to the model.


## 7.3.4 Interval estimates for new observations

Regression models are models for predicting a response variable given one or more predictor variables. We have seen how to obtain predictions (point estimates) of the response variable using predict for lm in Example 7.3 (see the predict.lm help topic). The predict method for lm also provides two types of interval estimates for the response:

a. Prediction intervals for new observations, for given values of the predictor variables.
b. Confidence intervals for the expected value of the response for given values of the predictors.

*Example 7.7 (Cherry Trees Model 3, cont.).*

To predict volume for new trees with given diameter and height, the **predict** method can be used. Store the diameter and height values for the new tree(s) in a data frame using the identical names as in the original model formula specified in the **lm** call. For example, to apply the Model 3 fit to obtain a point estimate for the volume of a new tree with diameter 16 in. and height 70 ft., we enter

```
> new = data.frame(Diam=16, Height=70)
> predict(M3, newdata=new)


       1
39.03278
```

The predicted volume of the new tree is 39.0 cubic feet. This estimate is about 10% lower than the prediction we obtained from Model 1, which used only diameter as a predictor.

To obtain a prediction interval or a confidence interval for the volume of the new tree, the **predict** method is used with an argument called **interval** specified. The confidence level is specified by the **level** argument, which is set at 0.95 by default. One can abbreviate the argument values. For a prediction interval specify **interval="pred"** and for a confidence interval use **interval="conf"**.

```
> predict(M3, newdata=new, interval="pred")

        fit      lwr      upr
1 39.03278 33.22013 44.84544
```

The prediction interval for volume of a randomly selected new tree of diameter 16 and height 70 is (33.2, 44.8) cubic feet. The confidence interval for the expected volume of all trees of diameter 16 and height 70 is obtained by

```
> predict(M3, newdata=new, interval="conf")

        fit      lwr      upr
1 39.03278 36.84581 41.21975
```

so the confidence interval estimate for expected volume is (36.8, 41.2) cubic feet. The prediction interval is wider than the confidence interval because the prediction for a single new tree must take into account the variation about the mean and also the variation among all trees of this diameter and height.

To obtain point estimates or interval estimates for several new trees, one would store the new values in a data frame like our data frame **new**. For example, if we require confidence intervals for diameter 16, at a sequence of values of height 65 to 70, we can do the following.

```
> diameter = 16
> height = seq(65, 70, 1)
> new = data.frame(Diam=diameter, Height=height)
> predict(M3, newdata=new, interval="conf")
```

which produces the following estimates:

```
        fit      lwr      upr
1 37.15085 34.21855 40.08315
2 37.52724 34.75160 40.30287
3 37.90362 35.28150 40.52574
4 38.28001 35.80768 40.75234
5 38.65640 36.32942 40.98338
6 39.03278 36.84581 41.21975
```

## 7.4 Fitting a Regression Curve

In this section we discuss two examples for which we want to estimate a regression curve rather than a straight line relation between a response variable $Y$ and a single predictor $X$. In Example 7.8 the response variable is linearly related to the reciprocal of the predictor. In Example 7.9, we fit an exponential model.

*Example 7.8 (Massachusetts Lunatics Data).*

In Chapter 1, the *Massachusetts Lunatics* data[1] was introduced. These data are from an 1854 survey conducted by the Massachusetts Commission on Lunacy. We created the data file "lunatics.txt" from the table on the web. See Chapter 1 (Example 1.12, page 29) for a detailed explanation of how to import the data into R. We import the data into a data frame lunatics and attach it using

```
> lunatics = read.table("lunatics.txt", header=TRUE)
> attach(lunatics)
```

The data frame lunatics has 14 rows and six columns, corresponding to the following variables:

| Variable | Description |
|----------|-------------|
| COUNTY | Name of county |
| NBR | Number of lunatics, by county |
| DIST | Distance to nearest mental health center |
| POP | County population , 1950 (thousands) |
| PDEN | County population density per square mile |
| PHOME | Percent of lunatics cared for at home |

---

[1] Data and Story Library, http://lib.stat.cmu.edu/DASL/Datafiles/lunaticsdat.html

In this example we investigate the relationship between the percentage of patients cared for at home and distance to the nearest health center.

First we plot PHOME vs DIST to see if a linear relation is a plausible model, and print the sample correlation.

```
> plot(DIST, PHOME)
> cor(DIST, PHOME)
[1] 0.4124404
```

The sample correlation 0.41 measures the linear association between the two variables. The scatterplot in Figure 7.8(a) suggests that the relation between PHOME and DIST is nonlinear, perhaps more like a hyperbola. With this in mind, we create the variable RDIST, the reciprocal of distance, compute sample correlation, and plot PHOME vs RDIST.

```
> RDIST = 1/DIST
> plot(RDIST, PHOME)
> cor(RDIST, PHOME)
[1] -0.7577307
```

Here $|\mathtt{cor(RDIST,PHOME)}| > |\mathtt{cor(DIST,PHOME)}|$, indicating a stronger linear association between RDIST and PHOME than between the original variables DIST and PHOME. In Figure 7.8(b) a linear relation between PHOME and RDIST appears to be a plausible model. (The line on the plot is added below after fitting the model.)

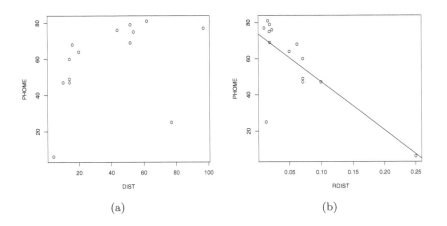

(a)                                                              (b)

**Fig. 7.8** Percent of lunatics cared for at home vs distance (a) and reciprocal of distance (b) in Example 7.8.

We fit the simple linear regression model

$$\mathrm{PHOME}_i = \beta_0 + \beta_1 \mathrm{RDIST}_i + \varepsilon_i, \qquad i = 1, \ldots, 14,$$

using the `lm` function. Typically we want to save the result in an object for further analysis.

```
> M = lm(PHOME ~ RDIST)
> M

Call:
lm(formula = PHOME ~ RDIST)

Coefficients:
(Intercept)          RDIST
      73.93         -266.32
```

The estimated regression line is $PHOME = 73.93 - 266.32\,RDIST$, and it can be added to the plot in Figure 7.8(b) with the `abline` function:

```
> abline(M)
```

Although data points for 13 of the 14 counties are close to the fitted line in Figure 7.8(b), there is one observation that is far from the line. We may also want to plot the fits for the original data. The fits are points on the curve

$$PHOME = \hat{\beta}_0 + \hat{\beta}_1 \frac{1}{DIST},$$

where $\hat{\beta}_0$ and $\hat{\beta}_1$ are the intercept and slope estimates stored in the `$coef` vector of our `lm` object. The plot in Figure 7.9 is obtained by

```
> plot(DIST, PHOME)
> curve(M$coef[1] + M$coef[2] / x, add=TRUE)
```

Again, we observe one observation that is far from the fitted curve. One also can observe that most of the observed data are above the fitted curve; the fitted model tends to underestimate the response.

A plot of residuals vs fits is produced by the command

```
> plot(M$fitted, M$resid, xlab="fitted", ylab="residuals")
```

We added a dashed horizontal line through 0 to the plot by

```
> abline(h=0, lty=2)
```

The plot is shown in Figure 7.10(a). On the residual plot we find that there is an outlier among the residuals at the lower right corner.

The `identify` function is helpful to identify which observation is the outlier. This function waits for the user to identify n points on the plot, and optionally labels the points. We want `n=1` to identify one point, and we specify an abbreviation for the `COUNTY` as the label.

```
> lab = abbreviate(COUNTY)
> identify(M$fitted.values, M$residuals, n=1, labels=lab)
[1] 13
```

The `identify` function returns the row number of the observation(s) identified on the plot. Row number 13 corresponds to NANTUCKET county, which

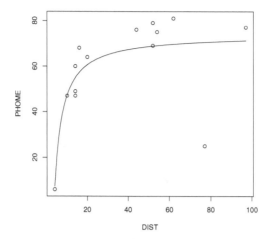

**Fig. 7.9** Predicted values of percentage of patients cared for at home in Example 7.8.

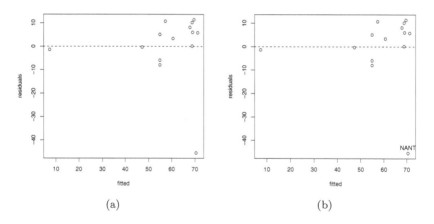

**Fig. 7.10** Fitted values vs residuals for the regression of PHOME on RDIST in Example 7.8. In (b), the outlier has been labeled as "NANT" (NANTUCKET county), after using the identify function.

is labeled on the plot as "NANT" where we clicked (See Figure 7.10(b).) We can extract this observation from the data set by

```
> lunatics[13, ]

     COUNTY NBR DIST  POP PDEN PHOME
13 NANTUCKET  12   77 1.74  179    25
```

According to the documentation provided with the data set on the DASL web site, Nantucket county is an offshore island, which may need to be taken into account in the model.

Finally, when the data frame is no longer required, it can be detached.

```
> detach(lunatics)
```

*Example 7.9 (Moore's Law).* In a recent interview,[2] Google's CEO Eric Schmidt discussed the future of the internet. According to Moore's Law, Schmidt said, "in 10 years every computer device you use will be 100 times cheaper or 100 times faster." Moore's Law, named for Intel co-founder Gordon Moore [34], states that the number of transistors on a chip (a measure of computing power) doubles every 24 months. In 1965 Moore predicted that the number of transistors would double every year, but in 1975 he modified that doubling time to every two years.

Moore's Law has been applied to various measurements of computing power. If the rate of growth in computing power is assumed constant, Moore's Law is the model

$$y = b_0 2^{b_1 t}, \qquad t \geq 0,$$

where $y$ is the measurement at time $t$, $b_0$ is the initial measurement, and $1/b_1$ is the time to double. That is, taking logarithms base 2 of both sides, we can write the model as

$$\log_2(y) = \log_2(b_0) + b_1 t, \qquad t \geq 0, \tag{7.2}$$

a linear model for logarithm of $y$ at time $t$.

In this example, we fit an exponential model to computer processor speed. The data file "CPUspeed.txt" contains the maximum Intel CPU speed vs time from 1994 through 2004. The variables are:

| Variable | Description |
|----------|-------------|
| year | calendar year |
| month | month |
| day | day |
| time | time in years |
| speed | Max IA-32 Speed (GHz) |
| log10speed | logarithm base 10 of speed |

The following code reads in the data from the file "CPUspeed.txt,"

```
> CPUspeed = read.table("CPUspeed.txt", header=TRUE)
```

and **head** displays the first few observations.

---

[2] http://firstdraftofhistory.theatlantic.com/analysis/internet_is_good.php

```
> head(CPUspeed)

  year month day      time speed log10speed
1 1994     3   7 1994.179 0.100 -1.0000000
2 1995     3  27 1995.233 0.120 -0.9208188
3 1995     6  12 1995.444 0.133 -0.8761484
4 1996     1   4 1996.008 0.166 -0.7798919
5 1996     6  10 1996.441 0.200 -0.6989700
6 1997     5   7 1997.347 0.300 -0.5228787
```

If Moore's Law holds (if an exponential model is correct), then we should expect that the logarithm of speed vs time follows an approximately linear trend. Here the base 2 logarithm is natural because of the base 2 logarithm in the proposed model (7.2), so we apply the change of base formula $\log_2(x) = \log_{10}(x)/\log_{10}(2)$. Since the earliest observation is in year 1994, we compute the time (years) measured in years since the start of 1994.

```
> years = CPUspeed$time - 1994
> speed = CPUspeed$speed
> log2speed = CPUspeed$log10speed / log10(2)
```

We construct scatterplots of speed vs time and $\log_2$(speed) vs time (where time is the time elapsed since 1994).

```
> plot(years, speed)
> plot(years, log2speed)
```

The plots are displayed in Figure 7.11(a) and 7.11(b). Figure 7.11(b) suggests that a linear association may exist between log2speed and years, so we fit a linear model using the lm function.

```
> L = lm(log2speed ~ years)
> print(L)

Call:
lm(formula = log2speed ~ years)

Coefficients:
(Intercept)        years
    -3.6581       0.5637
```

The fitted model is $\widehat{\ln y} = -3.6581 + 0.5637t$ or

$$\hat{y} = 2^{-3.6581+0.5637t} = 0.0792\,(2^{0.5637t}).\tag{7.3}$$

At time $t = 1/0.5637 = 1.774$ years, the predicted speed is $\hat{y} = 0.0792(2)$; thus, expected speed will double in an estimated 1.774 years. According to this model, CPU speeds are predicted to increase by a factor of $2^{0.5637(10)} \approx 50$ in 10 years (about 50 times faster, rather than 100 times faster as claimed in the interview).

To add the fitted regression curve to the plot in Figure 7.11(a) the curve function can be used with the exponential model (7.3).

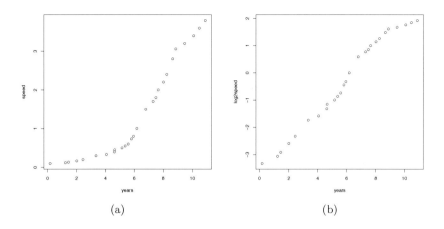

**Fig. 7.11** Plot of CPU speed vs time (a) and log2(CPU speed) vs time (b) in Example 7.9

```
> plot(years, speed)
> curve(2^(-3.6581 + 0.5637 * x), add=TRUE)
```

To add the fitted regression line to the plot in Figure 7.11(b) the `abline` function can be used.

```
> plot(years, log2speed)
> abline(L)
```

These two plots are displayed in Figures 7.12(a) and 7.12(b).

### Moore's Law: residual analysis

The fit of the model appears to be good from a visual inspection of the fitted line in Figure 7.11(b). The model adequacy can be investigated further through residual plots.

The residuals are observed errors $e_i = y_i - \hat{y}_i$, where $y_i$ is the observed response and $\hat{y}_i$ is the fitted value for observation $i$. The `lm` function returns an object containing residuals, fitted values, and other values. If we store the model rather than print it, we can access the residuals and other data returned by `lm`. In addition we can use available methods such as `summary`, `anova`, or `plot`.

The `plot` method for `lm` objects displays several residual plots. Using the argument `which=1:2` selects the first two plots.

```
> plot(L, which=1:2)
```

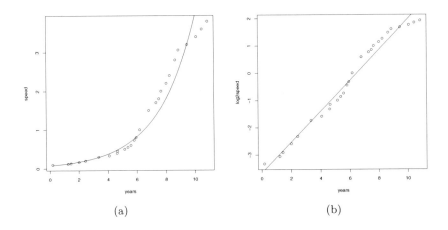

(a)                                    (b)

**Fig. 7.12** Plot of CPU speed vs time with fitted regression curve (a) and log2(CPU speed) vs time with fitted regression line (b) in Example 7.9

The two residual plots are shown in Figure 7.13. Figure 7.13(a) is a plot of residuals vs fitted values, with a curve added that has been fit to the data using a local regression "smoother" (see `lowess`). Figure 7.13(b) is a normal QQ plot of residuals.

An assumption for inference in regression is that the errors are independent and identically distributed (iid) as Normal$(0, \sigma^2)$, but the normal QQ plot suggests that the residuals are not normal. In Figure 7.13(a) it appears that the residuals are not iid.

Three larger residuals are identified on the plot of residuals vs fitted values (observations 16, 26, 27) and the same points are identified on the QQ plot. These are

```
> CPUspeed[c(16, 26, 27), ]
```

```
   year month day     time speed log10speed
16 2000    10  20 2000.802   1.5  0.1760913
26 2004     6  21 2004.471   3.6  0.5563025
27 2004    11  15 2004.873   3.8  0.5797836
```

Observations 26 and 27 are the two most recent, possibly indicating that the model is not a good fit for the near future.

The summary method produces additional information about the model fit. Suppose that one only needs the coefficient of determination $R^2$, rather than the complete output of `summary`. For simple linear regression extract `$r.squared` from the summary, and for multiple linear regression extract `$adj.r.squared` (adjusted $R^2$).

```
> summary(L)$r.squared
[1] 0.9770912
```

The coefficient of determination is 0.9770912; more than 97.7% of the total variation in the logarithm of speed is explained by the model.

On November 13, 2005 the maximum processor speed was 3.8 GHz. What does the fitted linear regression model predict for expected maximum speed at this time? The `predict` method for `lm` objects returns the predicted values for the data or new observations. The new observations should be stored in a data frame, using the same names for the predictor variables as in the model formula. In this case the data frame has a single variable, `years`. The fractional year up to November 13 should be expressed as a decimal. If we take the fractional year elapsed in 2005 to be 316.5/365, we have

```
> new = data.frame(years = 2005 + 316.5 / 365 - 1994)
> lyhat = predict(L, newdata=new)
> lyhat
       1
3.031005
```

Recall that the response variable in the fitted model is $\log_2(\text{speed})$, so the speed predicted by the model on November 13, 2005 is

```
> 2^lyhat
       1
8.173792
```

GHz and the error is 8.2 - 3.8 = 4.4 GHz. This illustrates the danger of extrapolation; note that our latest observation in `CPUspeed` is about one full year earlier, Nov. 15, 2004.

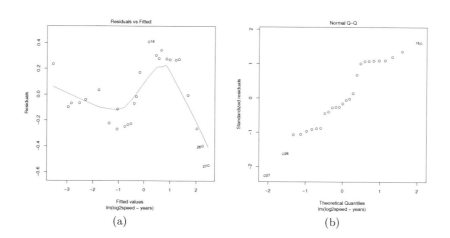

**Fig. 7.13** Residual plots from the fitted regression model, log2(CPU speed) vs time, in Example 7.9.

# Exercises

**7.1 (*mammals* data).** The `mammals` data set in the `MASS` package records brain size and body size for 62 different mammals. Fit a regression model to describe the relation between brain size and body size. Display a residual plot using the plot method for the result of the `lm` function. Which observation (which mammal) has the largest residual in your fitted model?

**7.2 (*mammals*, continued).** Refer to the `mammals` data in package `MASS`. Display a scatterplot of `log(brain)` vs `log(body)`. Fit a simple linear regression model to the transformed data. What is the equation of the fitted model? Display a fitted line plot and comment on the fit. Compare your results with results of Exercise 7.1.

**7.3 (*mammals* residuals).** Refer to Exercise 7.2. Display a plot of residuals vs fitted values and a normal-QQ plot of residuals. Do the residuals appear to be approximately normally distributed with constant variance?

**7.4 (*mammals* summary statistics).** Refer to Exercise 7.2. Use the `summary` function on the result of `lm` to display the summary statistics for the model. What is the estimate of the error variance? Find the coefficient of determination ($R^2$) and compare it to the square of the correlation between the response and predictor. Interpret the value of ($R^2$) as a measure of fit.

**7.5 (Hubble's Law).** In 1929 Edwin Hubble investigated the relationship between distance and velocity of celestial objects. Knowledge of this relationship might give clues as to how the universe was formed and what may happen in the future. Hubble's Law is is

$$\text{Recession Velocity} = H_0 \times \text{Distance},$$

where $H_0$ is Hubble's constant. This model is a straight line through the origin with slope $H_0$. Data that Hubble used to estimate the constant $H_0$ are given on the DASL web at `http://lib.stat.cmu.edu/DASL/Datafiles/Hubble.html`. Use the data to estimate Hubble's constant by simple linear regression.

**7.6 (*peanuts* data).** The data file "peanuts.txt" (Hand et al. [21]) records levels of a toxin in batches of peanuts. The data are the average level of aflatoxin $X$ in parts per billion, in 120 pounds of peanuts, and percentage of non-contaminated peanuts $Y$ in the batch. Use a simple linear regression model to predict $Y$ from $X$. Display a fitted line plot. Plot residuals, and comment on the adequacy of the model. Obtain a prediction of percentage of non-contaminated peanuts at levels 20, 40, 60, and 80 of aflatoxin.

**7.7 (*cars* data).** For the `cars` data in Example 7.1, compare the coefficient of determination $R^2$ for the two models (with and without intercept term in the model). Hint: Save the fitted model as L and use `summary(L)` to display $R^2$. Interpret the value of $R^2$ as a measure of the fit.

**7.8 (*cars* data, continued).** Refer to the `cars` data in Example 7.1. Create a new variable `speed2` equal to the square of `speed`. Then use `lm` to fit a quadratic model

$$dist = \beta_0 + \beta_1 speed + \beta_2 (speed)^2 + \varepsilon.$$

The corresponding model formula would be `dist ~ speed + speed2`. Use `curve` to add the estimated quadratic curve to the scatterplot of the data and comment on the fit. How does the fit of the model compare with the simple linear regression model of Example 7.1 and Exercise 7.7?

**7.9 (Cherry Tree data, quadratic regression model).** Refer to the Cherry Tree data in Example 7.3. Fit and analyze a quadratic regression model $y = b_0 + b_1 x + b_2 x^2$ for predicting volume $y$ given diameter $x$. Check the residual plots and summarize the results.

**7.10 (*lunatics* data).** Refer to the "lunatics" data in Example 7.8. Repeat the analysis, after deleting the two counties that are offshore islands, NAN-TUCKET and DUKES counties. Compare the estimates of slope and intercept with those obtained in Example 7.8. Construct the plots and analyze the residuals as in Example 7.8.

**7.11 (*twins* data).** Import the data file "twins.txt" using `read.table`. (The commands to read this data file are shown in the twins example in Section 3.3, page 85.) The variable `DLHRWAGE` is the difference (twin 1 minus twin 2) in the logarithm of hourly wage, given in dollars. The variable `HRWAGEL` is the hourly wage of twin 1. Fit and analyze a simple linear regression model to predict the difference `DLHRWAGE` given the *logarithm* of the hourly wage of twin 1.

# Chapter 8
# Analysis of Variance I

## 8.1 Introduction

Analysis of Variance (ANOVA) is a statistical procedure for comparing means of two or more populations. As the name suggests, ANOVA is a method for studying differences in means by analysis of the variance components in the model. In earlier chapters we have considered two sample location problems; for example, we compared the means of two groups using a two-sample $t$ test. Let us now consider a generalization to the multi-sample location problem, where we wish to compare the location parameters of two or more groups. One-way ANOVA handles a special case of this problem, testing for equal group means.

The following introductory example provides a context for reviewing terminology and statements about the models that follow.

*Example 8.1 (Eye color and flicker frequency).*
The data set "flicker.txt" measures 'critical flicker frequency' for 19 subjects with different eye colors.[1] From the description of the data on OzDASL [38] "an individual's critical flicker frequency is the highest frequency at which the flicker in a flickering light source can be detected. At frequencies above the critical frequency, the light source appears to be continuous even though it is actually flickering." Is critical flicker frequency related to eye color?

This data set has 19 observations and two variables, *Colour* (eye color) and *Flicker* (critical flicker frequency in cycles per second). The data is listed in Table 8.1.

Here we have a quantitative variable (Flicker) and a group variable (Colour). To formulate a model and hypothesis to test, let us ask a more specific question: Does the mean critical flicker frequency differ by eye color? The dependent variable or ***response*** is Flicker. The explanatory variable is

---

[1] See http://www.statsci.org/data/general/flicker.html, for the description and data; source [44].

| (a) | | |
|-----|-----|-----|
| Brown | Green | Blue |
| 26.8 | 26.4 | 25.7 |
| 27.9 | 24.2 | 27.2 |
| 23.7 | 28.0 | 29.9 |
| 25.0 | 26.9 | 28.5 |
| 26.3 | 29.1 | 29.4 |
| 24.8 |      | 28.3 |
| 25.7 |      |      |
| 24.5 |      |      |

| (b) | |
|-----|-----|
| Colour | Flicker |
| Brown | 26.8 |
| Brown | 27.9 |
| Brown | 23.7 |
| Brown | 25.0 |
| Brown | 26.3 |
| Brown | 24.8 |
| Brown | 25.7 |
| Brown | 24.5 |
| Green | 26.4 |
| Green | 24.2 |
| Green | 28.0 |
| Green | 26.9 |
| Green | 29.1 |
| Blue | 25.7 |
| Blue | 27.2 |
| Blue | 29.9 |
| Blue | 28.5 |
| Blue | 29.4 |
| Blue | 28.3 |

**Table 8.1** Two data layouts shown for the "Eye Color and Flicker Frequency" data. Table (a) on the left is in a spreadsheet-like layout (unstacked). Table (b) on the right is in stacked format. The response is *Flicker* (critical flicker frequency) and the factor is *Colour* (eye color). The factor *Colour* has three levels (Brown, Green, Blue).

the group variable, called the **treatment** or **factor**, and it has three **levels** in this example (Brown, Green, Blue). The null hypothesis is

$H_0$ : the mean response is equal for all groups.

A more general null hypothesis is that the location parameter of each group is equal; for example, one could test whether the medians differ by group. Even more general is a null hypothesis that the distribution of the response variable is identical for all groups. Either of these more general hypotheses can be tested using methods other than ANOVA, but ANOVA tests the hypothesis of equal means.

### 8.1.1 Data entry for one-way ANOVA

Data for one-way ANOVA is usually formatted in one of two ways. When data is given in a spreadsheet or a table, with each group's response in a different column, it is sometimes called "unstacked" or "wide" format. The same data could be organized in two columns, with the response variable in one column and the group variable (the factor) in another column; this format is sometimes called "stacked" or "long" format. See Table 8.1 for a comparison of the two formats.

For analysis in most software packages, including R, the data should be entered in *stacked* format (see e.g. Table 8.1(b)); that is, two columns corresponding to the two variables (the response and the group variable).[2]

First we illustrate the basic data entry steps for the *flicker* data, and do some exploratory data analysis.

*Example 8.2.* The information provided with the "flicker" data source indicates that there are two variables, Colour (eye color) and Flicker (critical flicker frequency). The data is available at `http://www.statsci.org/data/general/flicker.txt`, and it is in stacked format, so `read.table` is a simple way to enter this data. We need to set `header=TRUE` to indicate that the first row contains variable names.

```
> flicker = read.table(file=
+     "http://www.statsci.org/data/general/flicker.txt",
+     header=TRUE)
```

Alternately, assuming that the data has been saved into a local file "flicker.txt" and is located in the current working directory,

```
flicker = read.table("flicker.txt", header=TRUE)
```

is an easy way to enter the data.

If the data is not available to read from a connection using `read.table`, or if it is not in the correct format, then one could prepare a text file formatted like Table 8.1(b).

### Factors

In R the data will be entered into a data frame, and in order to proceed with ANOVA, the group variable should be a `factor`. A `factor` is a special type of object in R, used to describe data that can have a finite number of values; for example, gender, marital status, and eye color are variables that are typically of the factor type. The different values that a factor can take are called **levels**.

Let us verify that the group variable for eye color is a `factor` and display some information about it.

```
> is.factor(flicker$Colour)
[1] TRUE
> levels(flicker$Colour)
[1] "Blue"  "Brown" "Green"
```

At a glance factors may appear to be character vectors, but this is misleading. Using `unclass` we can reveal what information is stored in this factor:

---

[2] Example 8.9 shows how to convert from unstacked to stacked format.

```
> unclass(flicker$Colour)
 [1] 2 2 2 2 2 2 2 2 2 3 3 3 3 3 3 1 1 1 1 1 1
attr(,"levels")
[1] "Blue"  "Brown" "Green"
```

Interestingly we see that the levels are stored as integers and the characters are simply labels for the levels.

Notice that we did nothing special above to convert `flicker$Colour` to a factor – how did `read.table` automatically create `Colour` as a factor rather than a character vector when creating the data frame? There is an option in R (`stringsAsFactors`) that controls whether character vectors should automatically be converted to factors in a data frame. The 'factory-fresh' default is `TRUE`, but this can be changed by the user.

$R_x$ **8.1** *The "automatic" conversion of character vectors to factors in data frames obviously will not work for data files where factors have been encoded with numbers. In this case it is up to the user to convert the group variable to a factor. This can be accomplished using* **as.factor** *or* **factor**; *see e.g. Example 8.8. Also see Chapter 9 for examples.*

### 8.1.2 Preliminary data analysis

Let us now continue with data analysis for the *flicker* data.

*Example 8.3 (Eye color and flicker frequency, cont.).* It will be more convenient to attach the data frame so that we can refer to the variables directly:

```
> attach(flicker)
```

Side-by-side or parallel boxplots provide a graphical summary of location and dispersion. For boxplots we supply a formula with the response variable on the left and the group variable on the right.

```
> boxplot(Flicker ~ Colour, ylab = "Flicker")
```

The boxplots are shown Figure 8.1(a).

(Try the command `plot(Colour, Flicker)`. Notice that `plot` displays side-by-side boxplots in this case. The `plot` default behavior depends on the type of data to be plotted.)

A one-dimensional scatterplot or dotplots of the data can be displayed by the `stripchart` function.

```
> stripchart(Flicker ~ Colour, vertical=TRUE)
```

The plot is shown in Figure 8.1(b). The plots in Figures 8.1(a)-8.1(b) suggest that the samples are similar with respect to variance, but with possibly different centers (different location). The blue and brown means are farther

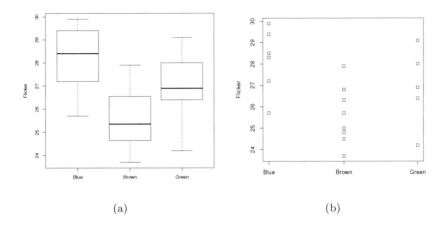

(a)    (b)

**Fig. 8.1** Box plots and dotplots of critical flicker frequency (cycles per minute) by eye color, in Example 8.1.

apart than the blue and green or brown and green, but it is not easy to tell from the plots if these differences are significant. Let us compare the means and standard deviations of the three groups.

A table of means can be obtained using the by command

```
by(Flicker, Colour, FUN=mean)
```

but we would like to customize the function to get both means and standard deviations. To do this, we define a function, meansd, that computes both mean and standard deviation, then supply this function name in the FUN argument of by.

```
> meansd = function(x) c(mean=mean(x), sd=sd(x))
> by(Flicker, Colour, FUN=meansd)
Colour: Blue
     mean         sd
28.166667   1.527962
--------------------------------------------------------
Colour: Brown
     mean         sd
25.587500   1.365323
--------------------------------------------------------
Colour: Green
     mean         sd
26.920000   1.843095
```

The standard deviations are close for the three groups, which we observed from the boxplots. However, the means of the three groups are also close; are there *significant* differences among the means?

## 8.2 One-way ANOVA

In a one-way ANOVA there is one treatment or factor $A$ with $a$ levels, and the null hypothesis is

$$H_0 : \mu_1 = \cdots = \mu_a,$$

where $\mu_j$ is the mean of the population identified by the $j^{th}$ level of the factor. The alternative hypothesis is that at least two of the means differ.

Sample variance is an average of squared deviations from the sample mean. The numerator of the sample variance is called the *corrected* sum of squares (SS), or sum of squared error about the mean. ANOVA partitions the total sum of squared error of the response (SS.total) into two components, the between-sample error (SST) and within-sample error (SSE). (See the appendix of this chapter for more details.)

If there are no differences among groups, then the mean squared error for between-sample and within-sample both estimate the same parameter. On the other hand, if there are differences among group means, then the expected between-sample mean squared error is larger (it includes the variation between groups) than when there are no differences. ANOVA tests the null hypothesis of equal treatment means by computing a ratio (the $F$ statistic) of the between and within mean squared errors. Details of the formulas and calculations are illustrated in Example 8.10.

There are several R functions available for one-way ANOVA calculations. The `oneway.test` function displays a summary of a hypothesis test of equal means, and is of course limited to a one-way analysis. The `lm` function fits linear models including one-way ANOVA models. The `aov` function is designed to fit ANOVA models, and it in turn calls `lm`. Methods to obtain the hypothesis tests and other analyses are available for both `lm` and `aov`. These functions are illustrated in the following sections.

### 8.2.1 ANOVA F test using `oneway.test`

*Example 8.4 (Eye color and flicker, cont.).* In this example, the null hypothesis is that the mean critical flicker frequency is equal for the three eye colors brown, green, and blue.

The R `oneway.test` function is a simple way to obtain the one-way analysis of variance. It is valid if certain conditions are satisfied: the errors are $NID(0, \sigma^2)$, where "NID" is an abbreviation for "normally distributed and independent." The symbol $\sigma^2$ is the error variance, and stating that errors are $NID(0, \sigma^2)$ also means that the variance of random error is equal for all groups. In the `oneway.test` function, there is an adjustment made (to degrees of freedom for error) for unequal variance. Normal error distribution can be checked using Normal-QQ plots of residuals (see e.g. Figure 8.2(b)).

```
> oneway.test(Flicker ~ Colour)

    One-way analysis of means (not assuming equal variances)

data:  Flicker and Colour
F = 5.0505, num df = 2.000, denom df = 8.926,
    p-value = 0.03412
```

Here the denominator degrees of freedom is non-integer, because error degrees of freedom have been adjusted to account for unequal variances. Large values of $F$ support the alternative. The test statistic is $F = 5.0505$ and the $p$-value $= 0.03412$ of the test is the area under the $F(2, 8.926)$ curve to the right of the point $F = 5.0505$. To test for equal means assuming equal variances, set var.equal=TRUE:

```
> oneway.test(Flicker ~ Colour, var.equal=TRUE)

        One-way analysis of means

data:  Flicker and Colour
F = 4.8023, num df = 2, denom df = 16, p-value = 0.02325
```

The $F$ statistic and denominator degrees of freedom are different than above, but in this example our conclusion to reject $H_0$ at 5% significance will not change.

The result of oneway.test is simply a brief report of the ANOVA $F$ test. This function does not return the ANOVA table, residuals, or fitted values. Although we have learned that there are some significant differences among the group means, we have no estimates (from oneway.test) for the means of groups, and no information about which group means differ.

The examples in Section 8.2.3 and 8.3 illustrate how to obtain a more detailed analysis using lm and aov functions and their methods.

## 8.2.2 One-way ANOVA model

In order to understand and interpret more detailed results of a one-way ANOVA, such as parameter estimates, it is important to understand how the model is specified in our particular software.

Let us denote the response variable by $Y$ so that $y_{ij}$ denotes the $i$-th observation in the $j$-th sample. Our model for critical flicker frequency by eye color could be written as

$$y_{ij} = \mu_j + \varepsilon_{ij}, \qquad i = 1, \ldots, n_j, \quad j = 1, 2, 3,$$

called the *means model*. The random error variable $\varepsilon_{ij}$ is assumed to have mean zero, and common variance $\sigma^2$, so $E[Y_{ij}] = \mu_j$, $j = 1, 2, 3$. The estimates for $\mu_1, \mu_2, \mu_3$ are the three group means by eye color $(\bar{y}_1, \bar{y}_2, \bar{y}_3)$. Residuals are the errors $y_{ij} - \bar{y}_j$. An estimate for $\sigma^2$ is the residual mean squared error. In general, the means model for one-way ANOVA with $a$ groups has $a + 1$ parameters: $\mu_1, \dots, \mu_a, \sigma^2$.

The *effects model* for a one-way ANOVA is

$$y_{ij} = \mu + \tau_j + \varepsilon_{ij}, \qquad i = 1, \dots, n_j, \quad j = 1, \dots, a, \qquad (8.1)$$

with the same assumptions for error. In the effects model, $\tau_j$ is a parameter that depends on group $j$, called the **treatment effect**. In this model $E[Y_{ij}] = \mu + \tau_j$. The null hypothesis of equal treatment means can be stated as $H_0 : \tau_j = 0$ for all $j$ (all groups have identical means equal to $\mu$).

In the effects model there are $a + 2$ parameters, so some restrictions are necessary. One of the $a + 2$ parameters must be a function of the other $a + 1$ parameters, so we need to place some constraints on the parameters of the effects model. If we take $\mu = 0$ then we have the means model with $\mu_j = \tau_j$. For one-way ANOVA in R the effects model is specified and the restriction is to set $\tau_1 = 0$. With this restriction, model (8.1) for three groups has $a + 1 = 4$ parameters: $\mu, \tau_2, \tau_3, \sigma^2$, the same number of parameters as the three group means model.

## 8.2.3 ANOVA using lm or aov

In Example 8.1, for the `boxplot` and `stripchart` functions we have used a *formula* that has the general form

$$\text{response} \sim \text{group}$$

and we specify a one-way ANOVA model in the same way. The R functions that implement ANOVA are `lm` (the same function that we used for linear regression models) and `aov` (analysis of variance). If the right hand side of the formula in the call to `lm` is a single numeric variable, the formula specifies a simple linear regression model; if the right hand side of the formula is a factor, the formula specifies a one-way ANOVA model.

The `aov` function fits an analysis of variance model by calling `lm`. For one-way ANOVA either function `lm` or `aov` obtains the same fitted model, but the displayed results are in a different format. See Section 8.3 for an example using `aov`.

### 8.2.4 Fitting the model

*Example 8.5 (Eye color and flicker, cont.).*
    The ANOVA model for this example is estimated by
lm(Flicker $\sim$ Colour), and saved in the object $L$ for further analysis.

```
> L = lm(Flicker ~ Colour)
> L

Call:
lm(formula = Flicker ~ Colour)

Coefficients:
(Intercept)  ColourBrown  ColourGreen
     28.167       -2.579       -1.247
```

The coefficients are the least squares estimates of the parameters $\mu, \tau_2, \tau_3$ in
(8.1). Now compare this table of coefficients returned by lm with the table of
means on page 203 computed with the by command. The intercept is equal
to the sample mean (28.166667) for blue eyes, but the other two coefficients
correspond to the differences between the sample means (25.5875-28.166667,
and (26.92-28.166667). Here $\hat{\mu}_j = \hat{\mu} + \hat{\tau}_j$ and the coefficients are $\hat{\mu}, \hat{\tau}_2, \hat{\tau}_3$, respectively (recall that the constraint is $\tau_1 = 0$).
    The fitted values $\hat{y}$ are the group means, $\bar{y}_j$. The fitted values are stored in
L$fitted.values, which can be abbreviated to L$fit. The predict method
for lm is another way to obtain the fitted values.

```
> predict(L)
        1        2        3        4        5        6        7
 25.58750 25.58750 25.58750 25.58750 25.58750 25.58750 25.58750
        8        9       10       11       12       13       14
 25.58750 26.92000 26.92000 26.92000 26.92000 26.92000 28.16667
       15       16       17       18       19
 28.16667 28.16667 28.16667 28.16667 28.16667
```

Here we see that the fitted values are indeed the group sample means for the
three eye colors. Comparing with the table of means on page 203, we can
observe that the predicted response corresponds to the group sample mean
for each observation.

### 8.2.5 Tables of means or estimated effects

The aov function can also be used to fit a one-way ANOVA model, and aov
has useful methods such as model.tables below, and TukeyHSD, which we
will use in Section 8.3.2. The table of means contains the grand mean, the
group means, and the number of replicates for each group.

```
> M = aov(Flicker ~ Colour)
> model.tables(M, type="means")
```

```
Tables of means
Grand mean

26.75263

 Colour
     Blue Brown Green
     28.17 25.59 26.92
rep   6.00  8.00  5.00
```

Another useful table is the table of effects, which is the type of table that model.tables displays by default.

```
> model.tables(M)

Tables of effects

 Colour
     Blue   Brown   Green
     1.414  -1.165  0.1674
rep  6.000   8.000  5.0000
```

The table of effects contains the estimates $\overline{y}_j - \overline{y}$ (the differences between the group means and the grand mean) and the number of replicates for each group.

## 8.2.6 ANOVA Table

The ANOVA table is obtained by the anova method applied to the lm object or summary method applied to the aov object. We might prefer to suppress printing of significance stars, which can be set through the options function as shown.

```
> options(show.signif.stars=FALSE)
> anova(L)

Analysis of Variance Table

Response: Flicker
          Df Sum Sq Mean Sq F value  Pr(>F)
Colour     2 22.997 11.4986  4.8023 0.02325
Residuals 16 38.310  2.3944
```

From the ANOVA table, the residual mean squared error (MSE) is 2.3944 with 16 degrees of freedom. This is the least squares estimate $\hat{\sigma}^2$ of error variance. (Observe that the $F$ statistic, degrees of freedom for error, and the $p$-value match the result of oneway.test with equal variance assumption on page 205.) For a valid $F$ test it is necessary that the samples are independent, and that the errors $\varepsilon_{ij}$ are independent and identically distributed normal variables with mean 0 and constant variance $\sigma^2$. If these conditions are satisfied then the $F$ test is significant at the 5% level (the $p$-value 0.02325 is less than 0.05), indicating that there are some differences in mean critical flicker frequency among eye color groups.

With such small sample sizes, it is difficult to assess normality and constant variance assumptions, but residual plots may help identify any severe departure from these assumptions. The fitted values of the model are the predicted response $\hat{y}_{ij} = \hat{\mu} + \hat{\tau}_j = \overline{y}_{.j}$. The residual for observation $y_{ij}$ is $e_{ij} = y_{ij} - \hat{y}_{ij}$. The fitted model stored in object L contains the residuals in variables L$residuals and the fitted values in L$fitted.values. A plot of residuals vs fitted values and a Normal-QQ plot is obtained by:

```
> #plot residuals vs fits
> plot(L$fit, L$res)
> abline(h=0)    #add horizontal line through 0
>
> #Normal-QQ plot of residuals with reference line
> qqnorm(L$res)
> qqline(L$res)
```

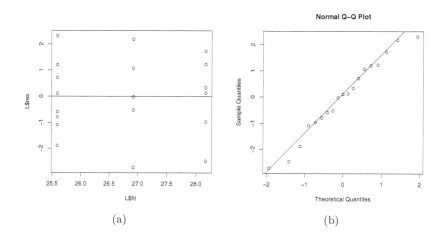

(a)                                         (b)

**Fig. 8.2** Residual plots for the one-way ANOVA model in Example 8.4.

The plots are shown in Figures 8.2(a) and 8.2(b). The residuals should be approximately symmetric about zero and have approximately equal variance. In the Normal-QQ plot the residuals should lie approximately along the reference line. The plots do not reveal any severe departure from the assumptions for this model.

## 8.3 Comparison of Treatment Means

Suppose that the null hypothesis of equal treatment means is rejected in a one-way ANOVA model, as in Example 8.4. Then we concluded that some differences exist among treatment means, but have not determined what these differences are. In this case it may be of interest to proceed with *post hoc* analysis, to find significant differences in means. For a one-way ANOVA with $a$ treatment means there are $\binom{a}{2} = a(a-1)/2$ different pairs of means. There are many methods available to conduct this analysis, called *multiple comparison methods*. Two commonly applied multiple comparison procedures are Tukey's test and the Fisher Least Significant Difference (LSD) method.

### 8.3.1 Fisher Least Significant Difference (LSD)

This procedure tests each of the hypotheses $H_0 : \mu_i = \mu_j$ vs $H_1 : \mu_i \neq \mu_j$, $1 \leq i < j \leq a$, using a two-sample $t$ test. The sample sizes are $n_j, j = 1,\ldots,a$ and the total sample size is $n_1 + \cdots + n_a = N$.

Under the assumptions of our model, errors are iid and normally distributed with common variance $\sigma^2$. The least squares estimate of $\sigma^2$ is the residual mean squared error MSE, with $N - a$ degrees of freedom. The parameter estimates are the sample means $\bar{y}_j$ and $Var(\bar{y}_j) = \sigma^2/n_j$. Thus the two-sample $t$ statistic for testing $H_0 : \mu_i = \mu_j$ is

$$T = \frac{\bar{y}_i - \bar{y}_j}{\sqrt{MSE\left(\frac{1}{n_i} + \frac{1}{n_j}\right)}}.$$

For a two-sided $t$ test, the pair of means $\mu_i$ and $\mu_j$ are significantly different if

$$|\bar{y}_i - \bar{y}_j| > t_{1-\alpha/2,N-a}\sqrt{MSE\left(\frac{1}{n_i} + \frac{1}{n_j}\right)},$$

where $t_{1-\alpha/2,N-a}$ is the $100(1-\alpha/2)$ percentile of the $t$ distribution with $N - a$ degrees of freedom. The *least significant difference* is

$$LSD = t_{1-\alpha/2,N-a}\sqrt{MSE\left(\frac{1}{n_i} + \frac{1}{n_j}\right)}.$$

If the design is balanced (all sample sizes are equal to $n$) then

$$LSD = t_{1-\alpha/2,a(n-1)}\sqrt{\frac{2\,MSE}{n}}.$$

*Example 8.6 (LSD method).* The LSD method can be applied to determine which pairs of means differ in Example 8.4. Recall that we have rejected the null hypothesis of equal means at 5% significance. Here $N = 19$ and $a = 3$. The sample means are 28.16667, 25.58750, and 26.92000, for blue, brown, and green eyes, respectively. There are three different pairs of means so there are three hypothesis tests.

The estimate of $\sigma^2$ is $MSE = 2.3944$ with 16 degrees of freedom. We can compute all pairwise differences between means using the outer function.

```
> MSE = 2.3944
> t97.5 = qt(.975, df=16)          #97.5th percentile of t
> n = table(Colour)                #sample sizes
> means = by(Flicker, Colour, mean) #treatment means

> outer(means, means, "-")
      Colour
Colour        Blue     Brown      Green
  Blue    0.000000 2.579167   1.246667
  Brown  -2.579167 0.000000  -1.332500
  Green  -1.246667 1.332500   0.000000
```

In the balanced design the same LSD would apply to all pairs. In this example sample sizes differ, but we can also use outer to compute all of the LSD values:

```
> t97.5 * sqrt(MSE * outer(1/n, 1/n, "+"))
      Colour
Colour       Blue     Brown     Green
  Blue   1.893888 1.771570 1.986326
  Brown  1.771570 1.640155 1.870064
  Green  1.986326 1.870064 2.074650
```

From the results above we can summarize:

```
               diff      LSD
Blue-Brown    2.5792   1.7716
Blue-Green    1.2467   1.9863
Brown-Green  -1.3325   1.8701
```

and conclude that mean critical flicker frequency for blue eyes is significantly higher than the mean for brown eyes, and no other pair is significantly different at 5%.

*Appropriate application of the LSD method:*

The multiple $t$ tests or $t$ intervals in the LSD method introduce the problem that our risk of rejecting a null hypothesis when in fact it is true (type 1 error) is increased. The Fisher LSD method usually refers to the situation where the comparisons are done *only after a significant ANOVA F test*. This restriction tends to avoid the problem of greatly inflated type 1 error rate in practice. See e.g. Carmer and Swanson [11].

An easy way to summarize all pairwise $t$-tests is provided by the function `pairwise.t.test`. The $p$-values of the test are reported in the table. The default method applies a two-sided test for equal means with pooled standard deviation, and the $p$-values are adjusted to control the type 1 error rate.

```
> pairwise.t.test(Flicker, Colour)

        Pairwise comparisons using t tests with pooled SD

data:  Flicker and Colour

      Blue  Brown
Brown 0.021 -
Green 0.301 0.301

P value adjustment method: holm
```

Of the $p$-values reported by `pairwise.t.test`, only the pair (blue, brown) have significantly different means at $\alpha = 0.05$.

### 8.3.2 Tukey's multiple comparison method

Tukey's method is based on the distance between the largest and smallest treatment means, called the range. The *studentized range statistic* is this difference divided by the estimated standard error:

$$q = \frac{\overline{y}_{\max} - \overline{y}_{\min}}{\sqrt{MSE/n}}.$$

The distribution of $q$ depends on the number of treatment means and the degrees of freedom for error. To apply Tukey's test, compute the critical difference

$$w = q_\alpha(a, df)\sqrt{\frac{MSE}{n}},$$

where $q_\alpha(a, df)$ is the $100(1 - \alpha)$ percentile of $q$, $a$ is the number of treatments, $n$ is the common sample size, and $df$ is the degrees of freedom for error. Conclude that the pair of means $(\mu_i, \mu_j)$ are significantly different at level $\alpha$ if $|\overline{y}_i - \overline{y}_j| > w$. (In the old days, one would refer to a table of critical values of $q$ to compute the critical difference. With R one can use the `qtukey` function for the percentiles of $q$.)

For balanced designs (equal sample sizes) the confidence level of a $(1 - \alpha)$ Tukey interval is exactly equal to $1 - \alpha$. This means that the test procedure controls type 1 error at level $\alpha$.

The modification for unequal sample size is the Tukey-Kramer method. The critical differences are

$$w = q_\alpha(a, df) \sqrt{\frac{MSE}{2} \left( \frac{1}{n_i} + \frac{1}{n_j} \right)}, \qquad i \neq j.$$

Tukey's multiple comparison procedure is sometimes called Tukey's Honest Significant Difference. It is implemented in the R function TukeyHSD.

*Example 8.7 (Tukey's multiple comparison procedure).* In Example 8.4 we have $a = 3$ treatment means and there are $df = 16$ degrees of freedom for error. Here we have unequal sample sizes so we apply the Tukey-Kramer method. The range statistic is 2.579167 (difference for blue and brown eyes). The qtukey function computes $q_{.05}(a = 3, df = 16)$ as

```
> qtukey(.95, nmeans=3, df=16)
[1] 3.649139
```

thus the critical differences at $\alpha = 0.05$ are

$$w = q_\alpha(a, df) \sqrt{\frac{MSE}{2} \left( \frac{1}{n_i} + \frac{1}{n_j} \right)} = 3.649139 \sqrt{\frac{2.3944}{2} \left( \frac{1}{n_i} + \frac{1}{n_j} \right)},$$

giving $w = 2.156$ (blue vs brown), $w = 2.418$ (blue vs green) and $w = 2.276$ (brown vs green). Refer to the table of differences on page 211. Only the first pair (blue vs brown) can be declared significantly different at $\alpha = 0.05$. This agrees with our results from the LSD method in Example 8.6.

One could also compute Tukey confidence intervals based on the studentized range statistic. For example, the 95% confidence interval for $\mu_2 - \mu_1$ is given by

$$\overline{y}_{brown} - \overline{y}_{blue} \pm w = -2.579 \pm 2.156 = (-4.735, -0.423).$$

Similarly one could compute confidence intervals for each difference $\mu_i - \mu_j$.

We can obtain the same result with much less effort using the TukeyHSD function. This function expects a fitted analysis of variance model as input, but the model should be fitted using the aov (analysis of variance) function. Actually, aov calls lm. First let us compare these two methods for one-way ANOVA.

```
> L = lm(Flicker ~ Colour)
> anova(L)
Analysis of Variance Table

Response: Flicker
          Df Sum Sq Mean Sq F value  Pr(>F)
Colour     2 22.997 11.4986  4.8023 0.02325
Residuals 16 38.310  2.3944

> M = aov(Flicker ~ Colour)
> M
Call:
   aov(formula = Flicker ~ Colour)
```

```
Terms:
                Colour Residuals
Sum of Squares  22.99729  38.31008
Deg. of Freedom        2        16

Residual standard error: 1.547378
Estimated effects may be unbalanced
```

We see that the two ANOVA tables are equivalent, although `aov` does not print the $F$ statistic and $p$-value. The `summary` method for `aov` objects produces the familiar ANOVA table.

```
> summary(M)
          Df Sum Sq Mean Sq F value  Pr(>F)
Colour     2 22.997 11.4986  4.8023 0.02325
Residuals 16 38.310  2.3944
```

Finally, we can get Tukey HSD confidence intervals by

```
> TukeyHSD(M)

  Tukey multiple comparisons of means
    95% family-wise confidence level

Fit: aov(formula = Flicker ~ Colour)

$Colour
                  diff        lwr       upr     p adj
Brown-Blue  -2.579167 -4.7354973 -0.422836 0.0183579
Green-Blue  -1.246667 -3.6643959  1.171063 0.3994319
Green-Brown  1.332500 -0.9437168  3.608717 0.3124225
```

We can refer to the $p$ value in the rightmost column to determine whether the difference in means is significantly different from 0. Only the first pair of means has $p$-value less than 0.05. Alternately we can look at the lower and upper confidence limits and check whether 0 is contained in the interval (`lwr`, `upr`). If 0 is inside the interval then the pair of means is not significantly different at 5%. Notice that we have obtained the same confidence interval for 'Brown-Blue' above by computing it from the Tukey-Kramer formula (see page 213).

The `TukeyHSD` function returns an object and there is a plot method for this object that displays a plot of the confidence intervals. We simply type

```
> plot(TukeyHSD(M))
```

and the plot of confidence intervals is displayed as shown in Figure 8.3. From the plot it is easy to see that only the first pair (top) are significantly different because only the first interval misses 0.

**95% family−wise confidence level**

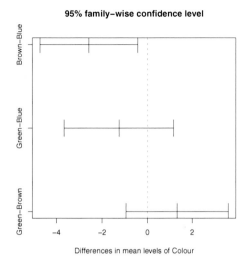

**Fig. 8.3** Tukey Honest Significant Difference confidence intervals in Example 8.7.

## 8.4 A Statistical Reference Dataset from NIST

*Example 8.8 (SiRstv from NIST).*

The data set "SiRstv" is one of the NIST Statistical Reference Datasets for analysis of variance. The description and the data are linked from the web page[3].

The data are measurements of bulk resistivity of silicon wafers made at NIST with 5 probing instruments on each of 5 days. The factor or treatment is the instrument, and there are $n = 5$ replicates for each instrument. The statistical model is

$$y_{ij} = \mu + \tau_j + \varepsilon_{ij}, \qquad i = 1, \ldots, 5, \quad j = 1, \ldots, 5,$$

where $y_{ij}$ is the $i$-th replicate for the $j$-th instrument. Recall that the constraint imposed by R model fitting is $\tau_1 = 0$.

The data is available in the two-column (stacked) format, but there is a description above the data. We copied the data portion and saved it in a text file "SiRstv.txt". The file contains the group variable **Instrument** (encoded with numbers 1 to 5) and **Resistance**, the response variable. Assuming that the data file is located in the current working directory, the data can be entered as follows.

```
> dat = read.table("SiRstv.txt", header=TRUE)
> head(dat)
```

---

[3] http://www.itl.nist.gov/div898/strd/anova/SiRstv.html

```
     Instrument Resistance
1            1    196.3052
2            1    196.1240
3            1    196.1890
4            1    196.2569
5            1    196.3403
6            2    196.3042
```

Here the factor levels are numbers, so the variable `Instrument` would not have been converted to a factor when `read.table` created the data frame. The first step to prepare the data for analysis is to convert `Instrument` to a factor.

```
> #Instrument is not a factor
> is.factor(dat$Instrument)
[1] FALSE

> #convert Instrument to factor
> dat$Instrument = as.factor(dat$Instrument)

> attach(dat)
> str(dat)        #check our result
'data.frame':   25 obs. of  2 variables:
 $ Instrument: Factor w/ 5 levels "1","2","3","4",..
 $ Resistance: num   196 196 196 196 196 ...
```

We display boxplots and dotplots of the data, shown in Figures 8.4(a) and 8.4(b), and compute a table of means and standard deviations.

```
> boxplot(Resistance ~ Instrument)
> stripchart(Resistance ~ Instrument, vertical=TRUE)
> by(Resistance, Instrument, FUN=function(x) c(mean(x), sd(x)))
Instrument: 1
[1] 196.2430800   0.0874733
---------------------------------------------
Instrument: 2
[1] 196.2443000   0.1379750
---------------------------------------------
Instrument: 3
[1] 196.16702000    0.09372413
---------------------------------------------
Instrument: 4
[1] 196.1481400   0.1042267
---------------------------------------------
Instrument: 5
[1] 196.14324000    0.08844797
```

$R_x$ **8.2** *In the `by` function above, the function is defined 'inline.' Compare this version with the application of `by` on page 203, where the function is defined outside of the `by` function. Try both versions on a data set, to see that the two versions produce the same result.*

The means and standard deviations appear to differ in the boxplots (Figure 8.4(a)), but in the table of means and standard deviations the means do not differ by more than two standard deviations.

Next we fit the model using `aov` and use the `summary` method to display the ANOVA table. A set of residual plots is easily displayed by the `plot` method.

```
> L = aov(Resistance ~ Instrument)
>
> summary(L)
            Df    Sum Sq  Mean Sq F value Pr(>F)
Instrument   4 0.051146 0.012787  1.1805 0.3494
Residuals   20 0.216637 0.010832

> par(mfrow=c(2, 2)) #4 plots on one screen
> plot(L)             #residual plots
```

NIST publishes certified results for the ANOVA table, displayed on the same web page as the data. We compared our result above to the certified values and they agree with the certified values rounded to the number of digits reported above. The $F$ statistic is not significant, so we cannot conclude that there are differences in mean resistivity measurements among the five instruments used. The residual plots are shown in Figure 8.5. These plots do not suggest severe departures from model assumptions.

The conclusion of the ANOVA $F$ test is also evident in the plot of confidence intervals using Tukey's HSD comparisons.

```
> par(mfrow=c(1, 1)) #restore display to plot full window
> plot(TukeyHSD(L))  #plot the Tukey CI's
```

There are $\binom{5}{2} = 10$ pairs of means, so a table can be difficult to interpret, but the plot in Figure 8.6 gives a nice visual comparison that is easy to interpret; all of the confidence intervals contain 0, so none of the pairs are significantly different at 5%.

$\mathbf{R_x}$ 8.3 *The web page for the SiRstv dataset indicates that the data starts on line 61, so we can use* `read.table` *to read directly from the web page if we skip the first 60 lines. The command would be*

```
file1 = "http://www.itl.nist.gov/div898/strd/anova/SiRstv.dat"
dat = read.table(file1, skip=60)
```

*When reading a file from a web page, the url is typically a long string, so our code is usually more readable if we assign the string to a file name as shown above. The name "*`file`*" is an R function so we assigned the value to* `file1`.

## 8.5 Stacking Data

The next example illustrates data entry for data that is provided in a table, similar to a spreadsheet layout.

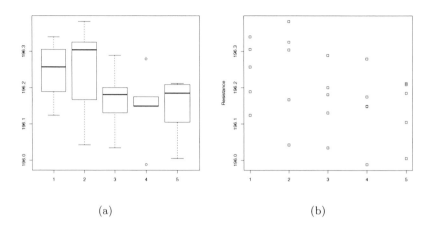

(a)                                                                        (b)

**Fig. 8.4** Box plots and dotplots of Resistance by Instrument, in Example 8.8.

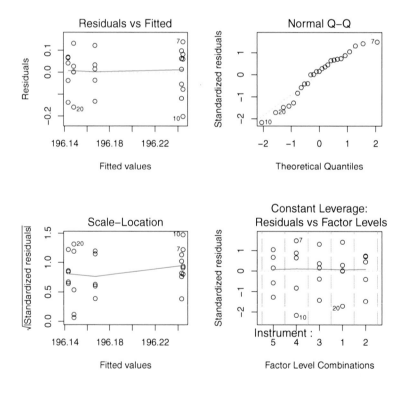

**Fig. 8.5** Residual plots for the NIST SiRstv data in Example 8.8.

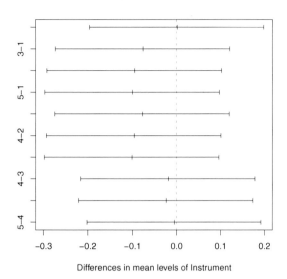

**95% family–wise confidence level**

Differences in mean levels of Instrument

**Fig. 8.6** Tukey HSD confidence intervals for differences in mean resistivity of instruments for the NIST SiRstv data in Example 8.8.

*Example 8.9 (Cancer survival times).* The data set "PATIENT.DAT" contains the survival times of cancer patients with advanced cancer of the stomach, bronchus, colon, ovary, or breast, whose treatment included supplemental ascorbate.[4] Do survival times differ by the type of cancer (organ)?

In order to analyze the data in a one-way layout, the data entry task is somewhat more complicated than above, because the data are not stacked. The data set can be opened with a plain text editor or with a spreadsheet to learn about the format of the data file. Comparing the data file with the description [21, p. 255], we see that the data file is organized in columns of different lengths corresponding to stomach, bronchus, colon, ovary, and breast cancers, respectively, but there is no header containing the labels. The data file is shown in Table 8.2.

Here `read.table` would expect each row to have an equal number of entries, so `read.table` will stop with an error at line 7. For example, if "PATIENT.DAT" is located in the current working directory,

```
> read.table("PATIENT.DAT")
Error in scan(file, what, nmax, sep, dec, ...
   line 7 did not have 5 elements
```

---

[4] Source: Hand et al. [21], from Cameron and Pauling [6].

**Table 8.2** In Example 8.2, the data file "PATIENT.DAT" containing survival times of cancer patients is organized in columns of different lengths corresponding to the affected organ: stomach, bronchus, colon, ovary, and breast cancers, respectively. The entries in each row are separated by tab characters at every fourth character. In this file there is no header containing the column labels.

```
124 81   248 1234      1235
42   461 377 89    24
25   20  189 201 1581
45   450 1843      356 1166
412 246 180 2970     40
51   166 537 456 727
1112      63  519      3808
46   64  455      791
103 155 406      1804
876 859 365      3460
146 151 942      719
340 166 776
396 37   372
         223 163
         138 101
         72  20
         245 283
```

To avoid this error, we need to provide more details to `read.table` about the file format. This data is separated by tab characters, which can be specified using the `sep="\t"` argument.

```
> times = read.table("PATIENT.DAT", sep="\t")
> names(times) = c("stomach","bronchus","colon","ovary","breast")
```

Here we used `names` to assign column names. Alternately, the column names can be specified using the `col.names` argument in `read.table`. The data in `times` is shown in Table 8.3 on page 221.

To stack the data one can use the `stack` function. The result of `stack` will retain the `NA` values.

```
> times1 = stack(times)
> names(times1)
[1] "values" "ind"
```

The result of `stack` generates a 'long' data frame with the numeric values of the response named `values` and the factor or group variable named `ind`. Let us rename the response and factor as `time` and `organ`. Also, now that the data is in stacked format, the `NA` values are not needed; `na.omit` removes the rows with `NA`.

```
> names(times1) = c("time", "organ")
> times1 = na.omit(times1)
```

**Table 8.3** Survival times of cancer patients from the file "PATIENT.DAT" after reading it into the data frame `times` in Example 8.2.

```
> times
   stomach bronchus colon ovary breast
1      124       81   248  1234   1235
2       42      461   377    89     24
3       25       20   189   201   1581
4       45      450  1843   356   1166
5      412      246   180  2970     40
6       51      166   537   456    727
7     1112       63   519    NA   3808
8       46       64   455    NA    791
9      103      155   406    NA   1804
10     876      859   365    NA   3460
11     146      151   942    NA    719
12     340      166   776    NA     NA
13     396       37   372    NA     NA
14      NA      223   163    NA     NA
15      NA      138   101    NA     NA
16      NA       72    20    NA     NA
17      NA      245   283    NA     NA
```

The beginning and end (`head` and `tail`) of the resulting data frame is shown below.

```
> head(times1, 3)
  time   organ
1  124 stomach
2   42 stomach
3   25 stomach
> tail(times1, 3)
   time  organ
77 1804 breast
78 3460 breast
79  719 breast
```

Next we verify that our group variable `organ` is a `factor`, and display a summary.

```
> is.factor(times1$organ)
[1] TRUE
> summary(times1)
      time             organ
 Min.   :  20.0   breast  :11
 1st Qu.: 102.5   bronchus:17
 Median : 265.5   colon   :17
 Mean   : 558.6   ovary   : 6
 3rd Qu.: 721.0   stomach :13
 Max.   :3808.0
```

$\mathbf{R_x}$ **8.4** *An alternate approach to data entry for tabular (spreadsheet-like) format is to use a spreadsheet to convert the file to tab-delimited (.txt) or comma-separated values (.csv) format. One advantage to using a spreadsheet for this type of data is that the data will be aligned in columns, which is often easier to read than in a data file such as the file "PATIENT.DAT" (Table 8.2). Once the file has been opened a header row can be inserted if desired for column names. Then after saving it in plain text tab-delimited or comma-delimited format,* `read.table` *or* `read.csv` *is convenient to input the data.*

```
> times = read.csv("PATIENT.csv")
> head(times)
  stomach bronchus colon ovary breast
1     124       81   248  1234   1235
2      42      461   377    89     24
3      25       20   189   201   1581
4      45      450  1843   356   1166
5     412      246   180  2970     40
6      51      166   537   456    727
```

Analysis of this data continues in Exercise 8.5.

## Exercises

**8.1 (Comparing average gas mileage).** A table of gas mileages on four new models of Japanese luxury cars is given in Larsen & Marx [30, Question 12.1.1], which is shown in Table 1.4, page 27. See Example 1.11 for a method of entering this data. Test if the four models (A, B, C, D) give the same gas mileage, on average, at $\alpha = 0.05$ significance.

**8.2 (Yields of plants).** The `PlantGrowth` data is an R dataset that contains results from an experiment on plant growth. The yield of a plant is measured by the dried weight of the plant. The experiment recorded yields of plants for a control group and two different treatments. After a preliminary exploratory data analysis, use one-way ANOVA to analyze the differences in mean yield for the three groups. Start with the exploratory data analysis, and also check model assumptions. What conclusions, if any, can be inferred from this sample data?

**8.3 (Differences in iris by species).** The `iris` data has 50 observations of four measurements for each of three species of iris: setosa, versicolor, and virginica. We are interested in possible differences in the sepal length of iris among the three species. Perform a preliminary analysis as in Example 8.3.

Write the effects model for a one-way ANOVA. What are the unknown parameters? Next fit a one-way ANOVA model for `Sepal.Length` by `Species` using `lm`. Display the ANOVA table. What are the parameter estimates?

**8.4 (Checking model assumptions).** Refer to your results from Exercise 8.3. What are the assumptions required for inference? Analyze the residuals of the model to assess whether there is a serious departure from any of these assumptions. How can you check for normality of the error variable?

**8.5 (Cancer survival data).** The cancer survival data "PATIENT.DAT" was introduced in Example 8.9. Start with the exploratory data analysis, and check for NID error model assumptions. Consider a transformation of the data if the assumptions for error are not satisfied. Complete a one-way ANOVA to determine whether mean survival times differ by organ. If there are significant differences, follow up with appropriate multiple comparisons to determine which means differ and describe how they differ.

**8.6 (Comparing waste at different manufacturing plants).** The 'Waste Run-up' data ([28, p. 86], [12]) is available at the DASL web site. The data are weekly percentage waste of cloth by five different supplier plants of Levi-Strauss, relative to cutting from a computer pattern. The question here is whether the five supplier plants differ in waste run-up. The data has been saved in a text file "wasterunup.txt". The five columns correspond to the five different manufacturing plants. The number of values in each column differs and the empty positions are filled with the symbol *. Use `na.strings="*"` in `read.table` to convert these to `NA` and use `stack` to reformat the data in a one-way layout. (Also see Example 10.1.)

Display side-by-side (parallel) boxplots of the data. Are any outliers in the data? Construct an ANOVA table using `lm` or `aov`. Plot residuals vs fits, and also construct a Normal-QQ plot of residuals using `qqnorm` and `qqline`. Do the residuals appear to satisfy the iid Normal$(0, \sigma^2)$ assumptions required for valid inference?

**8.7 (Verifying ANOVA calculations).** Refer to the chapter appendix. Verify the calculations for the ANOVA table in Exercise 8.3, by computing the entries of the ANOVA table directly from the formulas for $SST$ and $SSE$, as shown in Example 8.10.

# 8.6 Chapter 8 Appendix: Exploring ANOVA calculations

In this appendix we discuss the computation of the ANOVA table. Recall that ANOVA partitions the total (corrected) sum of squared error into two components, the between-sample error (SST) and within-sample error (SSE). Recall that $a$ is the number of samples (number of levels of the factor) and $n_1, \ldots, n_a$ are the sample sizes at each level. The total sum of squared error is

$$SS.total = \sum_{j=1}^{a}\sum_{i=1}^{n_j}(y_{ij} - \bar{y})^2.$$

The degrees of freedom for SS.total is $N - 1$, where $N$ is the total number of observations. The within-sample error is

$$SSE = \sum_{j=1}^{a}(n_j - 1)s_j^2,$$

where $s_j^2$ is the sample variance of the $j$-th sample. The degrees of freedom associated with SSE is the sum of the degrees of freedom, $\sum_{j=1}^{a}(n_j - 1) = N - a$. The sum of squares for treatments is

$$SST = \sum_{j=1}^{a}\sum_{i=1}^{n_j}(\bar{y}_j - \bar{y})^2 = \sum_{j=1}^{a}n_j(\bar{y}_j - \bar{y})^2.$$

The degrees of freedom for treatments are $a - 1$, so $df(SST) + df(SSE) = N - a = df(SS.total)$. The squared errors for treatments and error also sum to the total squared error; that is,

$$SS.total = SST + SSE.$$

The mean-squared error is the sum of squared error divided by degrees of freedom: $MSE = SSE/(N - a)$; $MST = SST/(a - 1)$.

The ANOVA $F$ statistic is the ratio

$$F = \frac{MST}{MSE} = \frac{SST/(a - 1)}{SSE/(N - a)}.$$

The $F$ ratio will be large if the average between-sample error $MSE$ is large relative to the mean within-sample error $MSE$. The expected values are

$$E[MST] = \sigma^2 + \frac{1}{a - 1}\sum_{j=1}^{a}n_j\tau_j^2,$$

$$E[MSE] = \sigma^2,$$

where $\tau_j$ are the effects in model (8.1). Under the null hypothesis of equal means, $E[MSE] = E[MST] = \sigma^2$, but under the alternative some $\tau_j \neq 0$ and $E[MST] > \sigma^2$. Thus large values of the ratio $F = MST/MSE$ support the alternative hypothesis that some $\tau_j \neq 0$. Under null hypothesis and NID errors, the ratio $F = MST/MSE$ follows an $F$ distribution with degrees of freedom $(a - 1, N - a)$, and the ANOVA $F$ test rejects the null hypothesis for large values of the $F$ statistic. The $p$-value of the test is $Pr(F(a - 1, N - a) > MST/MSE)$. For more details refer to e.g. Montgomery [33, Ch. 3].

*Example 8.10 (ANOVA calculations).* We have displayed an ANOVA table
for the flicker data in Example 8.4 using the `anova` method for `lm`. In this
example we illustrate the basic calculations behind the entries in the ANOVA
table. The preliminary steps to input the data are (see Example 8.1):

```
> flicker = read.table(file=
+     "http://www.statsci.org/data/general/flicker.txt",
+     header=TRUE)
> attach(flicker)
```

Then SS.total and SSE are found by:

```
> n = table(Colour)       #sample sizes
> a = length(n)           #number of levels
> N = sum(n)              #total sample size
> SS.total = (N - 1) * var(Flicker)
> vars = by(Flicker, Colour, var)    #within-sample variances
> SSE = sum(vars * (n - 1))
```

These calculations match the ANOVA table on page 208.

```
> print(c(a - 1, N - a, N - 1))      #degrees of freedom
[1]  2 16 18
> print(c(SS.total - SSE, SSE, SS.total)) #SST, SSE, SS.total
[1] 22.99729 38.31008 61.30737
```

Now let us verify that SST=SS.total-SSE for this example by computing SST
from the sample means.

```
> means = by(Flicker, Colour, mean) #treatment means
> grandmean = sum(Flicker) / N
> SST = sum(n * (means - grandmean)^2)
> print(SST)
[1] 22.99729
```

This calculation also agrees with the ANOVA table. Finally, the calculation
of the mean squared errors, $F$ statistic, and $p$-value in the ANOVA table:

```
> MST = SST / (a - 1)
> MSE = SSE / (N - a)
> statistic = MST / MSE
> p.value = pf(statistic, df1=a-1, df2=N-a, lower.tail=FALSE)
```

We used the $F$ cdf function `pf` with `lower.tail=FALSE` to compute the upper
tail probability for the $p$-value of the $F$ statistic. Then `list` is a convenient
way to display the results with labels:

```
> print(as.data.frame(
+     list(MST=MST, MSE=MSE, F=statistic, p=p.value)))
       MST      MSE        F         p
1 11.49864 2.39438 4.802346 0.02324895
```

Again these calculations match the ANOVA table on page 208.

# Chapter 9
# Analysis of Variance II

## 9.1 Introduction

Analysis of Variance (ANOVA) is a statistical procedure for comparing means of two or more populations. In Chapter 8 we considered one-way ANOVA models, which help analyze differences in the mean response corresponding to the levels of a single group variable or factor. In this chapter we consider randomized block designs and two-way ANOVA models. Randomized block designs model the effects of a single group variable or factor while controlling for another source of variation using blocks. Two-way ANOVA models explain differences in the mean response corresponding to the levels of two group variables (factors) and their possible interaction.

## 9.2 Randomized Block Designs

In this section we focus on *randomized block designs*, which model the effect of a treatment or factor where the samples are not independent, but matched by a variable called a block variable.

In experiments some sources of variation are known and controllable. In Example 9.1 the *experimental units* are students who have taken a series of exams. The objective is to compare the student performance on different types of exams. There is of course, inherent variability among students, but the effect of this variability on the statistical analysis of exam scores can be controlled by introducing a variable to identify the student (the block variable). Other sources of variation may be unknown and uncontrollable. Randomization is a design technique that guards against this type of variation.

*Example 9.1 (Exam scores).* The `scor` data is provided in the `bootstrap` package [31]. It contains exam scores in five subjects for 88 students. The five exams are

| Name | Description | Type |
|------|-------------|------|
| mec | mechanics | closed book |
| vec | vectors | closed book |
| alg | algebra | open book |
| ana | analysis | open book |
| sta | statistics | open book |

The layout of the `scor` data is shown in Table 9.1. In this example, a one-way analysis is not appropriate. A one-way ANOVA analyzes a completely randomized design, where the samples are assumed to be independent. If the exam topics (mechanics, vectors, algebra, analysis, statistics) are considered to be the treatments, the exam scores for each topic are not independent samples because the same 88 students took each exam.

**Table 9.1** Layout of the `scor` data in Example 9.1. There are 88 rows corresponding to 88 students, and five columns corresponding to five different exam scores for each student.

| mec | vec | alg | ana | sta |
|-----|-----|-----|-----|-----|
| 77 | 82 | 67 | 67 | 81 |
| 63 | 78 | 80 | 70 | 81 |
| 75 | 73 | 71 | 66 | 81 |
| 55 | 72 | 63 | 70 | 68 |
| 63 | 63 | 65 | 70 | 63 |
| 53 | 61 | 72 | 64 | 73 |
| ⋮ | | | | |
| 0 | 40 | 21 | 9 | 14 |

To use this data set, install the `bootstrap` package (see the example in Chapter 1, page 33). Once the `bootstrap` package is installed, load it using `library`. Then the data frame `scor` is available, and we can print the first part of the data frame using the `head` function.

```
> library(bootstrap)
> head(scor)

  mec vec alg ana sta
1  77  82  67  67  81
2  63  78  80  70  81
3  75  73  71  66  81
4  55  72  63  70  68
5  63  63  65  70  63
6  53  61  72  64  73
```

A table of means by exam can be obtained by `sapply`:

```
> sapply(scor, mean, data=scor)

     mec      vec      alg      ana      sta
38.95455 50.59091 50.60227 46.68182 42.30682
```

### 9.2.1 The randomized block model

Suppose that we want to compare the mean exam scores for different exams. Then the response is the exam score, and the treatment or factor is the topic of the exam. Let us assume that the scores are matched by student, so that each row of the data frame corresponds to an individual student's scores. Each row is called a block in this example, and the block variable is the student.

Let $y_{ij}$ denote the score for student $j$ on exam $i$, $i = 1,\ldots,5$, $j = 1,\ldots,88$. The model for this block design can be written as

$$y_{ij} = \mu + \tau_i + \beta_j + \varepsilon_{ij}, \tag{9.1}$$

where $\tau_i$ are the treatment effects, $\beta_j$ are the block effects, and $\varepsilon_{ij}$ are iid normally distributed errors with mean 0 and common variance $\sigma^2$.

The general formula for this type of model in R has the syntax:

$$\text{response} \sim \text{treatment} + \text{block}$$

*Example 9.2 (Exam scores, continued).*
   In order to proceed with ANOVA using R, we need to reshape the data so that the response $y_{ij}$ is one variable (one column in a data frame), the treatment label is a factor (another column), and the block label is a factor (also one column).

**R**x **9.1** *Another way to reshape the data is to enter it into a spreadsheet and "manually" reshape the data and create the factors. Save the data in plain text format (comma or tab delimited) and use* **read.table** *or* **read.csv** *to read it into an R data frame. See Example 9.3.*

   The **stack** function puts the response in a single column with the column names as a factor variable (the treatment). We then add the student ID factor (block variable) to the resulting data frame.

```
> scor.long = stack(scor)
> block = factor(rep(1:88, times=5))
> scor.long = data.frame(scor.long, block)
```

It is a good idea to check that the result is correct. The data set is now 440 lines long, so we only show the top and bottom.

```
> head(scor.long)  #top
  values ind block
1     77 mec    1
2     63 mec    2
3     75 mec    3
4     55 mec    4
5     63 mec    5
6     53 mec    6
> tail(scor.long)  #bottom
    values ind block
435     18 sta   83
436     17 sta   84
437     18 sta   85
438     21 sta   86
439     20 sta   87
440     14 sta   88
```

Next we rename the variables in scor.long

```
> names(scor.long) = c("score", "exam", "student")
```

and check the result using the structure function (str).

```
> str(scor.long)
'data.frame':   440 obs. of  3 variables:
 $ score  : num  77 63 75 55 63 53 51 59 62 64 ...
 $ exam   : Factor w/ 5 levels "alg","ana","mec",..: 3 3 3 3 3 3 3 3 ...
 $ student: Factor w/ 88 levels "1","2","3","4",..: 1 2 3 4 5 6 7 8 ...
```

## 9.2.2 Analysis of the randomized block model

Now the data is ready for analysis using lm or aov to fit the model (9.1) and anova or summary to display the ANOVA table. In this chapter we use the aov function to fit the model.

The main hypothesis of interest is that there is no difference among the mean exam scores for different exams. We may also test whether there are differences among students.

```
> L = aov(score ~ exam + student, data=scor.long)
> summary(L)
            Df Sum Sq Mean Sq F value    Pr(>F)
exam         4   9315 2328.72  21.201 1.163e-15 ***
student     87  58313  670.26   6.102 < 2.2e-16 ***
Residuals  348  38225  109.84
```

Recall that large values of the $F$ statistic support the alternative hypothesis. Here the $F$ statistics for both treatment (exam) and block (student) are very large (the $p$-values are very small) and both tests are significant. We can conclude that there are significant differences among the mean exam scores and significant differences between mean student scores.

The `model.tables` function is useful for producing a nicely formatted table of effect estimates.

```
> model.tables(L, cterms="exam")
Tables of effects

 exam
exam
   alg    ana    mec    sta    vec
 4.775  0.855 -6.873 -3.520  4.764
```

Here the effect estimates are the differences between the treatment (exam) means and the grand mean. This can be checked using `model.tables` to display a table of means:

```
> model.tables(L, cterms="exam", type="mean")
Tables of means
Grand mean

45.82727

 exam
exam
  alg   ana   mec   sta   vec
50.60 46.68 38.95 42.31 50.59
```

The Tukey Honest Significant Difference method for comparing means was introduced in Chapter 8. Using the `TukeyHSD` function with the fit from the `aov` function, it is easy to display confidence intervals for the differences in means. Here we specify that confidence intervals should be computed for only the first factor by `which=1`. The `TukeyHSD` function displays a table of 95% confidence intervals by default.

```
> CIs = TukeyHSD(L, which=1)
> CIs
  Tukey multiple comparisons of means
    95% family-wise confidence level

Fit: aov(formula = score ~ exam + student)

$exam
               diff         lwr         upr      p adj
ana-alg  -3.92045455  -8.2530400   0.41213094 0.0972395
mec-alg -11.64772727 -15.9803128  -7.31514179 0.0000000
sta-alg  -8.29545455 -12.6280400  -3.96286906 0.0000026
vec-alg  -0.01136364  -4.3439491   4.32122185 1.0000000
mec-ana  -7.72727273 -12.0598582  -3.39468725 0.0000152
sta-ana  -4.37500000  -8.7075855  -0.04241452 0.0464869
vec-ana   3.90909091  -0.4234946   8.24167639 0.0989287
sta-mec   3.35227273  -0.9803128   7.68485821 0.2131645
vec-mec  11.63636364   7.3037782  15.96894912 0.0000000
vec-sta   8.28409091   3.9515054  12.61667639 0.0000027
```

From the table displayed by the `TukeyHSD` function we can identify significant differences between several pairs of mean exam scores, but note that "vec-alg" and "sta-mec" are not significant. Also "ana-alg" and "vec-ana" are not significant at 5%. The two closed book exams (mechanics and vectors) are significantly different as are several pairs of the open book mean exam scores.

Optionally we can display the results graphically by: plot(CIs), but the vertical axis labels are easier to read if we specify horizontal labels by the las argument:

```
> plot(CIs, las=1)
```

The plot produced is displayed in Figure 9.1.

The confidence intervals from top to bottom correspond to the rows of the matrix returned by TukeyHSD, shown above. The differences in means are not significantly different at 5% if the confidence interval covers zero.

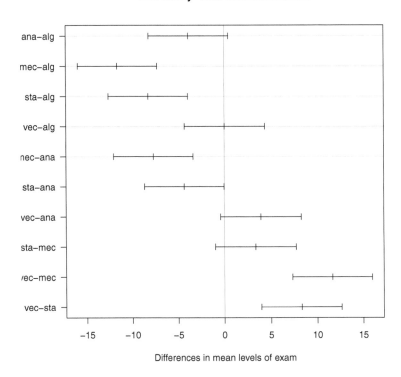

**Fig. 9.1** Tukey HSD confidence intervals for differences in mean exam scores in Example 9.2.

To check the assumptions for the distribution of the error term, it is helpful to plot residuals. Below we display the first two of several available plots, a scatterplot of residuals and a normal-QQ plot.

```
> plot(L, which=1:2)
```

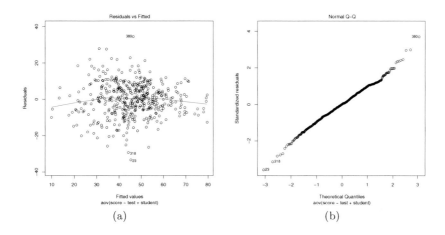

**Fig. 9.2** Residual plots for the two-way block model fit on `scor` data.

The Normal QQ plot in Figure 9.2(b) does not reveal a serious departure from the normality assumptions, although there are some outliers labeled on the QQ plot. An inspection of the data reveals a zero score and several very low scores in the 'vector' exam, for example. The plot of residuals vs fits in Figure 9.2(a) has a 'football' or elliptical shape, which may indicate nonconstant error variance. Exam scores are often percentages, which have nonconstant variance like proportions. Recall that the variance of a sample proportion is $p(1-p)/n$, where $p$ is the true population proportion, and the variance is largest when $p = 1/2$.

It is also interesting to look at either the profiles of students' scores or other types of plots to compare them. In this case, boxplots (Figure 9.3) are easier to interpret than line plots for comparing students.

```
> boxplot(score ~ student,
+   xlab="Student Number", ylab="Score", data=scor.long)
```

One thing revealed in the boxplots of scores by student is that the variability in student scores may depend on the ability of the student; students with high scores tend to have smaller variation in scores.

Some remedies for nonconstant variance include transformation of the response, using a different type of model, or a nonparametric method of analysis. A nonparametric approach based on randomization tests is covered in Chapter 10.

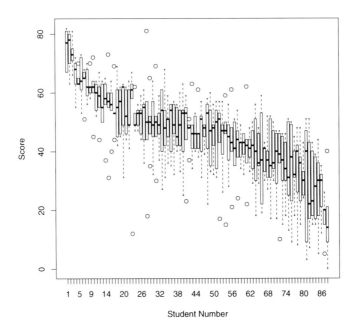

**Fig. 9.3** Boxplots of exam scores by student.

## 9.3 Two-way ANOVA

In a two-way factorial design there are two factors and there may be an interaction between the two factors.

*Example 9.3 (Poisons).* Table 9.2 gives the survival times in units of 10 hours for animals exposed to four different poisons. The data appears in Box, Hunter & Hunter [5, p. 228]. We converted this data to the stacked format by entering it into a spreadsheet "poison.csv" which can be read into R using

```
> poison = read.csv("poison.csv")
```

The `poison` data frame contains three variables labeled Time, Poison, Treatment. To get a quick summary of the structure of this data frame, we use `str` (structure).

```
> str(poison)
```

```
'data.frame':   48 obs. of  3 variables:
 $ Time     : num  0.31 0.45 0.46 0.43 0.36 0.29 0.4 0.23 0.22 0.21 ...
 $ Poison   : Factor w/ 3 levels "I","II","III": 1 1 1 1 2 2 2 2 3 3 ...
 $ Treatment: Factor w/ 4 levels "A","B","C","D": 1 1 1 1 1 1 1 1 1 1 ...
```

The `str` command shows that there are 48 observations and 3 variables: Time (numeric), and factors Poison and Treatment. Notice that characters in data frames are automatically converted into factor type variables.

**Table 9.2** Survival times in units of 10 hours for animals exposed to poisons I, II, III, and antidotes A, B, C, D.

|        | treatment |      |      |      |
|--------|------|------|------|------|
| poison | A    | B    | C    | D    |
| I      | 0.31 | 0.82 | 0.43 | 0.45 |
|        | 0.45 | 1.10 | 0.45 | 0.71 |
|        | 0.46 | 0.88 | 0.63 | 0.66 |
|        | 0.43 | 0.72 | 0.76 | 0.62 |
| II     | 0.36 | 0.92 | 0.44 | 0.56 |
|        | 0.20 | 0.61 | 0.35 | 1.02 |
|        | 0.40 | 0.49 | 0.31 | 0.71 |
|        | 0.23 | 1.24 | 0.40 | 0.38 |
| III    | 0.22 | 0.30 | 0.23 | 0.30 |
|        | 0.21 | 0.37 | 0.25 | 0.36 |
|        | 0.18 | 0.38 | 0.24 | 0.31 |
|        | 0.23 | 0.29 | 0.22 | 0.33 |

## 9.3.1 The two-way ANOVA model

The layout of the poison data (Table 9.2) corresponds to a $3 \times 4$ factorial design. It is replicated 4 times. Because there may be interaction between the type of poison and the type of antidote, an interaction term should be included in the model. The model can be specified as

$$y_{ijk} = \mu + \alpha_i + \beta_j + (\alpha\beta)_{ij} + \varepsilon_{ijk}, \tag{9.2}$$

where $\alpha_i$ are the poison type effects, $\beta_j$ are the antidote type effects, and $(\alpha\beta)_{ij}$ is the interaction effect for poison $i$ and antidote $j$. The index $k$ corresponds to the replicates so $y_{ijk}$ is survival time for the $k-th$ animal who received the $i-th$ poison and $j-th$ antidote, and errors $\varepsilon_{ijk}$ are iid and normally distributed with mean 0 and common variance $\sigma^2$. Interaction terms are specified in R model formulas using the colon operator. For example,

```
Time ~ Poison + Treatment + Poison:Treatment
```

is the model syntax corresponding to model (9.2). A shorter formula for the same model is specified by

```
Time ~ Poison * Treatment
```

## 9.3.2 Analysis of the two-way ANOVA model

The analysis of this data is in four parts outlined below.

1. Fit a two way model with interaction, display the ANOVA table, and test for interaction. If interaction is not significant, test for significant main effects.
2. Construct interaction plots, and identify any possible sources of interaction.
3. Use Tukey's multiple comparison method to compare group means. Compare the result with a multiple comparison by pairwise $t$-tests.
4. Analyze residuals of the model to investigate whether there are any apparent departures from the model assumptions. (This is left as an exercise.)

*Example 9.4 (Two-way ANOVA Step 1: Fit the model and display the ANOVA table).*

```
> L = aov(Time ~ Poison * Treatment, data = poison)
> anova(L)
Analysis of Variance Table

Response: Time
                 Df  Sum Sq Mean Sq F value    Pr(>F)
Poison            2 1.03708 0.51854 23.3314 3.176e-07 ***
Treatment         3 0.92012 0.30671 13.8000 3.792e-06 ***
Poison:Treatment  6 0.25027 0.04171  1.8768    0.1118
Residuals        36 0.80010 0.02222
```

The $F$ statistic for poison-treatment interaction is not significant at $\alpha = 0.05$ ($p$-value $= 0.1118$). The $F$ statistic for main effect of poison is significant ($p$-value $< .001$) and the $F$ statistic for main effect of treatment (antidote) is significant ($p$-value $< .001$).

*Example 9.5 (Two-way ANOVA Step 2: Interactions).*
Two versions of an interaction plot are generated below and shown in Figures 9.4(a) and 9.4(b). Since `interaction.plot` does not accept a **data** argument, we use **with** to specify that the variables are in the **poison** data frame.

```
> with(data=poison, expr={
+    interaction.plot(Poison, Treatment, response=Time)
+    interaction.plot(Treatment, Poison, response=Time)
+ })
```

Interaction plots are line plots of the group means for one factor vs another factor. If there is no interaction, the slopes of the line segments should be approximately equal; the line segments should be approximately parallel. In Figure 9.4(a), the lines corresponding to antidotes B and D have somewhat

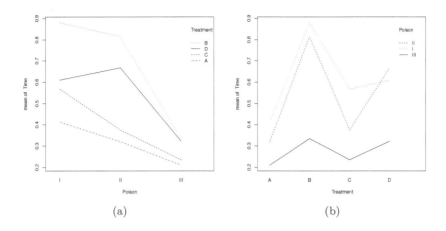

**Fig. 9.4** Interaction plots for `Poison` and `Treatment` in Example 9.5.

different slopes than the lines corresponding to antidotes A and C. In Figure 9.4(b), the lines corresponding to poison III have different slopes than poisons I and II. The ANOVA $F$ test for Poison-Treatment interaction, however, indicated that interactions are not significant at $\alpha = 0.05$.

*Example 9.6 (Two-way ANOVA (Step 3)).*

Step 3. Comparison of means

Tables of means for the main effects and interactions can be displayed using the `model.tables` function.

```
> model.tables(L, type="means")
Tables of means
Grand mean

0.4791667

 Poison
Poison
     I     II    III
0.6175 0.5444 0.2756

 Treatment
Treatment
     A      B      C      D
0.3142 0.6767 0.3925 0.5333
```

```
Poison:Treatment
     Treatment
Poison A      B       C       D
   I    0.4125 0.8800 0.5675 0.6100
   II   0.3200 0.8150 0.3750 0.6675
   III  0.2100 0.3350 0.2350 0.3225
```

Tables of effects can also be displayed using `model.tables(L)` (not shown). The larger means correspond to longer survival times. Smaller means correspond to more toxic poisons or less effective antidotes. However, from the table of means one cannot determine which poisons or antidotes are significantly better or worse. Multiple comparison procedures provide a formal test procedure to determine which poisons or antidotes are significantly better or worse than others.

Multiple comparison of means using Tukey's Honest Significant Difference method is provided in R by the `TukeyHSD` function. By default, `TukeyHSD` will compute comparisons for all factors in the model, including the interaction. Here the interaction was not significant, so we specify that we only want the comparisons for main effects by the `which` argument.

```
> TukeyHSD(L, which=c("Poison", "Treatment"))

  Tukey multiple comparisons of means
    95% family-wise confidence level

Fit: aov(formula = Time ~ Poison * Treatment)

$Poison
            diff        lwr         upr       p adj
II-I    -0.073125 -0.2019588  0.05570882 0.3580369
III-I   -0.341875 -0.4707088 -0.21304118 0.0000005
III-II  -0.268750 -0.3975838 -0.13991618 0.0000325

$Treatment
            diff        lwr          upr       p adj
B-A   0.36250000  0.19858517  0.52641483 0.0000046
C-A   0.07833333 -0.08558150  0.24224816 0.5769172
D-A   0.21916667  0.05525184  0.38308150 0.0050213
C-B  -0.28416667 -0.44808150 -0.12025184 0.0002320
D-B  -0.14333333 -0.30724816  0.02058150 0.1045211
D-C   0.14083333 -0.02308150  0.30474816 0.1136890
```

Results from Tukey's HSD procedure indicate that there are significant differences between mean survival times for Poisons (I, III) and (II, III) and there is a significant difference between mean survival times for antidotes (A, B), (A, D), and (B, C). The interpretation is that poison III is the most toxic, and antidote B is significantly more effective than A or C.

For a nice graphical summary of these results, there is a plot method

```
> plot(TukeyHSD(L, which=c("Poison", "Treatment")))
```

which plots confidence intervals for the differences (not shown).

*Remark 9.1.* When the interaction is significant, one should only compare the interaction means for each factor at a fixed level of the other factor. For example, if poison-treatment interaction was significant, we would compare the means for the antidotes on poison I, antidotes on poison II, and antidotes on poison III, separately. Comparisons on the overall poison means or the overall antidote means would not be meaningful.

Pairwise *t*-tests for poison and antidote are easily obtained by the `pairwise.t.test` function, as shown in Table 9.3.

**Table 9.3** Pairwise *t*-tests for the `poison` data.

```
> pairwise.t.test(poison$Time, poison$Poison)

        Pairwise comparisons using t tests with pooled SD

data:  Time and Poison

     I       II
II   0.3282  -
III  9.6e-05 0.0014

P value adjustment method: holm

> pairwise.t.test(poison$Time, poison$Treatment)

        Pairwise comparisons using t tests with pooled SD

data:  Time and Treatment

    A      B      C
B 0.0011 -      -
C 0.3831 0.0129 -
D 0.0707 0.3424 0.3424

P value adjustment method: holm
```

From the pairwise *t*-tests with Holm's *p*-value adjustment we can conclude that there are significant differences between mean survival times for poison types (I, III) and (II, III). Average survival time is longest for poisons I and II. Among antidotes, only the means of types (A, B) and (B, C) are significantly different at $\alpha = 0.05$. Antidote B means are significantly higher than A or C.

The conclusions of the pairwise *t*-tests agree with the Tukey method except that the pairwise *t*-test procedure did not find a significant difference between means of antidotes A and D that was significant using the Tukey method.

# Exercises

**9.1 (Rounding first base).** The data in "rounding.txt" gives the times required to round first base for 22 baseball players using three styles: rounding out, a narrow angle and a wide angle.[1] The goal is to determine if the method of rounding first base has a significant effect on times to round first base.

The data and the format of the data can be viewed using a text editor or a spreadsheet. With the data file in the current working directory, input the data using

```
rounding = read.table("rounding.txt", header=TRUE)
```

Check using the `str` function that the data is in stacked format with three variables: `time`, `method`, `player`, where `time` is numeric and `method` and `player` are factors.

Analyze the data assuming a randomized block design with time to round first base as the response, method of rounding first as the treatment, and player as the block variable. Plot residuals to check whether the assumptions for the distribution of random error appear to be valid.

**9.2 (Speed of light).** The `morley` data in R contains the classical data of Michaelson and Morley on the speed of light, recording five experiments of 20 consecutive runs each. The response is the speed of light measurement `Speed`. The experiment is `Expt` and the run is `Run`. See the documentation (`?morley`) and also `http://lib.stat.cmu.edu/DASL/Stories/SpeedofLight.html` for more details about the experiments and the data set.

Use the `str` function to check that there are 100 observations of the response `Speed`, `Expt`, and `Run`; all integer variables. Convert `Expt` and `Run` to factors using

```
morley$Expt = factor(morley$Expt)
morley$Run = factor(morley$Run)
```

Display a boxplot of `Speed` by `Expt`. Speed of light is a constant, so we see there are some problems because the measurements of speed do not seem to be consistent across the five experiments.

The data can be viewed as a randomized block experiment. What is the null hypothesis of interest? Analyze the data and residuals and summarize your conclusions.

**9.3 (Residual analysis for poison survival times).** Complete the analysis (Step 4) of the two-way factorial model for the poison survival times in Example 9.3. Prepare a plot of residuals against predicted values and any other plots that may help to assess the validity of the model and model assumptions.

---

[1] Source: Hollander and Wolfe, [25, Table 7.1, page 274].

**9.4 (Reciprocal transformation for poison survival times).** Refer to Exercise 9.3. The residual plots suggest a reciprocal transformation of the response (poison survival time), $z_{ijk} = 1/y_{ijk}$. Make this transformation, and repeat the entire analysis of Example 9.3 and Exercise 9.3. Summarize the analysis and discuss any differences between the original model and the current model. Are there any differences evident in the interaction plots?

# Chapter 10
# Randomization Tests

## 10.1 Introduction

If the model assumptions for ANOVA do not hold, then the ANOVA $F$ test is not necessarily valid for testing the hypothesis of equal means. However, one can compute an ANOVA table and an $F$ statistic; what is in doubt is whether the "$F$" ratio has an $F$ distribution.

A *randomization test* or *permutation test* provides a nonparametric approach based on the $F$ statistic that does not require that the test statistic ($F$) has an $F$ distribution. Permutation tests were introduced in the basic inference chapter (Section 6.4.4) for testing the two-sample hypothesis of equal means. Just as ANOVA generalizes the two-sample $t$-test test for equal means to $k \geq 2$ samples, the randomization tests discussed in this chapter generalize the two-sample permutation test discussed in Section 6.4.4. The main idea of a randomization (permutation) test is explained below in Section 10.3.

## 10.2 Exploring Data for One-way Analysis

Prior to applying formal methods of statistical inference in any data analysis problem, it is essential to explore the data with descriptive and graphical summaries. If the research question is to determine whether groups differ in location, then the preliminary analysis helps to determine whether a one factor model is reasonable or whether there may be one or more other variables that should be included in the model. In the exploratory analysis one can check informally whether certain parametric model assumptions hold, which helps to identify the type of analysis (parametric or nonparametric) that is most suitable for the data at hand.

*Example 10.1 (The 'Waste Run-up' data).* The 'Waste Run-up' data ([28, p. 86], [12]) is available at the DASL web site. The data are weekly percentage waste of cloth by five different supplier plants of Levi-Strauss, relative to cutting from a computer pattern. The question here is whether the *five supplier plants differ in waste run-up.*

The data has been saved in a text file "wasterunup.txt". The five columns correspond to the five different manufacturing plants. In the data file, the numbers of values in each column differ and the empty positions are filled with the symbol *.

To convert this data into the one-way layout for comparison of groups, we first need to read the data into R. We use read.table to read the text file into R. The special "*" character is specified by setting na.strings="*" in the arguments to read.table. Then the data can be stacked using the stack function, which places all observations into a single column and creates an index variable labeled with column names. The result is exactly what we need except for the NA values, which can be removed by na.omit on the result.

```
> waste = read.table(
+     file="wasterunup.txt",
+     header=TRUE, na.strings="*")
> head(waste)   #top of the data set

   PT1   PT2  PT3  PT4  PT5
1   1.2  16.4 12.1 11.5 24.0
2  10.1  -6.0  9.7 10.2 -3.7
3  -2.0 -11.6  7.4  3.8  8.2
4   1.5  -1.3 -2.1  8.3  9.2
5  -3.0   4.0 10.1  6.6 -9.3
6  -0.7  17.0  4.7 10.2  8.0

> waste = stack(waste) #stack the data, create group var.
> waste = na.omit(waste)
```

$\mathbf{R}_{\mathbf{x}}$ **10.1** *The argument* file="wasterunup.txt" *below assumes that the data file is located in the current working directory, which can be displayed using* getwd() *and changed using* setwd. *Alternately, one could supply a path name; for example, if data files are in a subdirectory of the current working directory called "Rxdatafiles" then the filename can be specified as* "./Rxdatafiles/wasterunup.txt".

Now after the stack and na.omit functions are applied, the resulting data set waste has two variables, values (the response, numeric) and ind (the plant, categorical), as the summary method shows.

```
> summary(waste)

     values          ind
 Min.   :-11.600   PT1:22
 1st Qu.:  2.550   PT2:22
 Median :  5.200   PT3:19
 Mean   :  6.977   PT4:19
 3rd Qu.:  9.950   PT5:13
 Max.   : 70.200
```

For the categorical variable `ind` (the `plant`), the summary function displays a frequency table corresponding to the sample sizes for the five plants. Another way to display the sample sizes is

```
> table(waste$ind)

PT1 PT2 PT3 PT4 PT5
 22  22  19  19  13
```

To change the variable names, we can use the `names` function. Below the name of the second variable "ind" is changed to "plant".

```
> names(waste)[2] = "plant"
> names(waste)

[1] "values" "plant"
```

The group means and five-number summaries can be displayed using the `summary` function. First we `attach` the `waste` data frame so that variables can be referenced directly by `values` and `plant` rather than `waste$values` and `waste$plant`. Alternately, one could use the `with` function (see Chapter 9 for an example).

```
> attach(waste)
>  by(values, plant, summary)
plant: PT1
   Min. 1st Qu.  Median    Mean 3rd Qu.    Max.
 -3.200  -0.450   1.950   4.523   3.150  42.700
----------------------------------------------------------------------
plant: PT2
   Min. 1st Qu.  Median    Mean 3rd Qu.    Max.
-11.600   3.850   6.150   8.832   8.875  70.200
----------------------------------------------------------------------
plant: PT3
   Min. 1st Qu.  Median    Mean 3rd Qu.    Max.
 -3.900   2.450   4.700   4.832   8.500  12.100
----------------------------------------------------------------------
plant: PT4
   Min. 1st Qu.  Median    Mean 3rd Qu.    Max.
  0.700   5.150   7.100   7.489  10.200  14.500
----------------------------------------------------------------------
plant: PT5
   Min. 1st Qu.  Median    Mean 3rd Qu.    Max.
  -9.30    8.00   11.30   10.38   16.80   24.00
```

Side-by-side boxplots and stripcharts are visual displays that convey the information in the five-number summaries as well as a measure of dispersion, the inter-quartile range (IQR). Comparing the range and IQR of the groups is one way to assess whether groups have approximately equal variances before fitting a model. Residuals of a fitted one-way model could also be compared in a boxplot. The R functions to obtain the boxplots and dotplots of the response variable are:

```
> boxplot(values ~ plant)
> stripchart(values ~ plant, vertical=TRUE)
```

The resulting plots are shown in Figure 10.1(a) and 10.1(b).

Alternately one could analyze the residuals from a fitted one-way model. The residuals are contained in the returned value of `lm`, which fits the one-way model. To check normality, a normal-QQ plot of residuals is useful. A plot of residuals vs fitted values, a normal-QQ plot of residuals, as well as other plots are obtained by `plot(L)`. Here we specified `which=1:2` in the `plot` command to obtain just the first two plots. Notice that the plot of residuals vs fitted values in Figure 10.1(c) is like the dotplot of the response variable in Figure 10.1(b), except that the horizontal axis has changed. (To obtain a boxplot of residuals, one could use `boxplot(L$resid ~ plant)`.)

```
> L <- lm(values ~ plant)
> plot(L, which=1:2)  #the first two residual plots
```

The plots in Figures 10.1(a) through 10.1(d) suggest that the data may have outliers and possibly that variances differ among the groups. Thus, the usual assumptions for distribution of errors required for valid inference with the ANOVA $F$-test may not hold in this case. Hence the ANOVA $F$ test is not reliable for inference in this situation. In the following sections we illustrate two nonparametric approaches that do not require the equal variance or normality of error conditions for valid inference.

## 10.3 Randomization Test for Location

The ANOVA $F$ statistic is a ratio that compares the between-sample and within-sample mean squared error. Whether or not the $F$ ratio has an $F$ distribution under the null hypothesis, we know that the large values of the $F$ statistic are more evidence that there are differences among group means than smaller $F$ values. Therefore even if our $F$ statistic does not have an $F$ distribution, we can apply an $F$ *statistic* (not an $F$ test) if we have a correct reference distribution under the null hypothesis.

The idea of the randomization test for a one-way analysis is simple. Suppose that there are no differences among treatments or groups. Then group labels assigned to the response could be randomly mixed-up without changing the sampling distribution of the test statistic. The randomization test generates a large number $R$ of samples of the observed response with randomly assigned group labels and computes an $F^*$ statistic for each sample. The collection of these $F^*$ statistics becomes our estimate for the distribution of our statistic $F$ when $H_0$ is true. Then our original observed $F$ statistic is compared to the replicates $F^*$. The estimated $p$-value is $\hat{p} = Pr(F \geq F^*) = \#(F^* \leq F)/(R+1)$. (We divide by $R+1$ because with the original $F$ we have a total of $R+1$ statistics.)

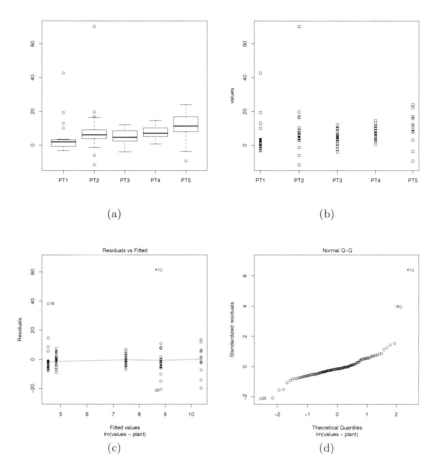

**Fig. 10.1** Boxplots and stripchart of "Waste Run-up" response variable, and plots of residuals from the fitted one-way analysis of variance model.

*Example 10.2 (Randomization test).* First let us write a function to perform the randomization test for an arbitrary response and group variable. The $F^*$ statistics are generated by the `replicate` function with these lines:

```
stats = replicate(R, {
  random.labels = sample(group)
  oneway.test(response ~ random.labels)$statistic})
```

The group labels are randomized by the permutation `sample(group)`. The $F^*$ statistics are computed for each sample by `oneway.test`. The resulting value of `stats` is a vector containing the R replicates $F^*$. Then we construct a function as a convenient user interface containing the simulation of $F^*$ and computation of the $p$-value for the test.

First let us see the display of oneway.test although we do not want to use it for inference. We will modify the $p$-value and the name of the test in our function but keep the format of the output so that the printed report is similar to oneway.test.

```
> oneway.test(values ~ plant)

        One-way analysis of means (not assuming equal
        variances)

data:  values and plant
F = 1.8143, num df = 4.000, denom df = 40.328,
p-value = 0.1449
```

The description of the test is the method. To avoid any confusion, we will modify test$method in the return value to indicate that we used a randomization test. If the number of replicates is not specified, the default value of R=199 replicates are generated.

```
rand.oneway = function(response, group, R=199) {
  test = oneway.test(response ~ group)
  observed.F <- test$statistic
  stats = replicate(R, {
    random.labels = sample(group)
    oneway.test(response ~ random.labels)$statistic})
  p = sum(stats >= observed.F) / (R+1)
  test$method = "Randomization test for equal means"
  test$p.value = p
  test
}
```

Now let's apply the randomization test to the *waste* data.

```
> rand.oneway(response=values, group=plant, R=999)

        Randomization test for equal means

data:  response and group
F = 1.8143, num df = 4.000, denom df = 40.328,
p-value = 0.128
```

Notice that the observed $F$ statistic is unchanged, but the $p$-value has changed. If we repeat this test, the reported $p$-value could vary slightly because we are generating random permutations. (In the randomization test the numerator and denominator degrees of freedom are not used.)

$\mathbf{R_x}$ **10.2** *Randomization tests for the one-way location problem and other problems are implemented in the* **coin** *package (conditional inference) [27]. If* **coin** *has been installed, then the following method can be used to obtain a randomization test based on a different statistic.*

```
> library(coin)
> oneway_test(values ~ plant, distribution=approximate(B=999))

        Approximative K-Sample Permutation Test

data:  values by
       plant (PT1, PT2, PT3, PT4, PT5)
maxT = 1.3338, p-value = 0.6306

> detach(package:coin)
```

*See the documentation in* **coin** *for details.*


## 10.4 Permutation Test for Correlation

Randomization tests can be implemented for a variety of inferential problems. Here we illustrate the use of a permutation test to provide a nonparametric alternative to the traditional test of a correlation coefficient.

*Example 10.3 (Website counts).* One of the authors maintains a website on a particular book and using Google Analytics, records the number of visits to this particular website on each day of the year. There are interesting patterns to these website "hits". As one might expect, there tend to be more hits on the website during weekdays than on weekends (Saturday and Sunday). Also since this book may be used for statistics courses, there appear to be a greater number of hits during the periods when classes are in session.

To explore the week by week patterns of these website visits, we collected the number of hits for 35 weeks from April through November, 2009. We read in the data file "webhits.txt" and construct a scatterplot of the number of website visits (variable Hits) against the week number (variable Week). As we look at the scatterplot in Figure 10.2, we see a gradual upward drift in the website hits for increasing week number. Is this pattern due to a "real" change in the counts over time, or is this pattern a simple byproduct of chance variation?

```
> web = read.table("webhits.txt", header=TRUE)
> plot(web$Week, web$Hits)
```

One can rephrase this question using a statistical test. Suppose the distribution of the webhits is unknown and we are uncomfortable making the assumption that the counts follow a normal distribution. In this case we want to apply a nonparametric test of association rather than the Pearson correlation test. Let us replace the website counts with their ranks, where a rank of 1 is assigned to the smallest count, a rank of 2 is assigned to the next-smallest count, and so on. Let $R_i$ denote the rank assigned to the $i$th website count. Our null hypothesis is that the sequence of website hit ranks, $R_1, ..., R_{35}$, is not associated with the week numbers 1 through 35.

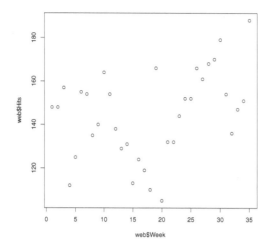

**Fig. 10.2** Scatterplot of website counts against week number.

One can measure the association by the correlation between the ranks and the week numbers

$$r_{obs} = \text{correlation}\{\text{Ranks, Week Number}\}.$$

The R function `cor.test` implements the Pearson correlation test as well as tests for two rank based measures of association, Spearman's $\rho$ and Kendall's $\tau$. Each of these tests the null hypothesis that the measure of association is zero. We can also implement the rank tests as permutation tests. We illustrate both methods for comparison. For Spearman's rank test we apply `cor.test` with `method="spearman"`.

```
> cor.test(web$Week, web$Hits, method="spearman")

        Spearman's rank correlation rho

data:  web$Week and web$Hits
S = 4842.713, p-value = 0.05945
alternative hypothesis: true rho is not equal to 0
sample estimates:
      rho
0.3217489

Warning message:
In cor.test.default(web$Week, web$Hits, method = "spearman") :
  Cannot compute exact p-values with ties
```

There are ties in the counts, so an exact $p$-value cannot be computed in this example. An approximate $p$ value is reported. (The warning message could be suppressed by adding the argument `exact=FALSE` in `cor.test`.) At the 5% significance level we cannot conclude that there is an association between `Week` and `Hits`.

## Permutation test approach

If the null hypothesis is true, that is, there is no relationship between the count numbers and the week numbers, then the distribution of the correlation should not change if we randomly permute the website counts (or equivalently the ranks).

This observation motivates a randomization or permutation test of our hypothesis of no association. We obtain a sampling distribution for the rank correlation statistic $r$ under the null hypothesis by considering the distributions of the correlation of (Rank(`Hits`), `Week`) over all possible permutations of the hit counts. There are $35! = 1.03 \times 10^{40}$ permutations. One can imagine computing the correlation $r$ of `Week` and permutation(Rank(`Hits`)) for each permutation, obtaining a null sampling distribution of $r$.

To see if the observed correlation value $r_{obs}$ is consistent with this null distribution, we compute the $p$-value, which is the probability that a correlation under the null distribution exceeds $r_{obs}$.

It is impractical to compute the exact $p$-value here since the number of permutations in the randomization distribution is too large, but it is practical to perform this permutation test by simulation. A function `rand.correlation` (below) is written to implement this procedure. There are three arguments; the vectors of paired data `x` and `y`, and the number of simulation replications `R`. In this function, we initially take the ranks of the two vectors and compute the observed rank correlation $r_{obs}$. We use the `replicate` function to repeatedly take a random permutation of the ranks of $y$ (using the `sample` function). The output of the function is the observed correlation and the $p$-value, which is the proportion of simulated correlations from the null distribution that exceed $r_{obs}$.

```
rand.correlation = function(x, y, R=199) {
    ranks.x = rank(x)
    ranks.y = rank(y)
    observed.r = cor(ranks.x, ranks.y)
    stats = replicate(R, {
      random.ranks = sample(ranks.y)
      cor(ranks.x, random.ranks)
    })
    p.value = sum(stats >= observed.r) / (R + 1)
    list(observed.r = observed.r, p.value = p.value)
}
```

We run this function for our example using 1000 simulation replications.

```
> rand.correlation(web$Week, web$Hits, R=1000)
$observed.r
[1] 0.3217489

$p.value
[1] 0.02997003
```

Here we observe that the rank correlation statistic $r_{obs}$ is 0.3217, and it is the equal to the Spearman rank correlation statistic. The computed $p$-value is 0.030, indicating that there is some evidence of a trend of the website hit counts over time.


## Exercises


**10.1 (*PlantGrowth* data).** The `PlantGrowth` data is an R dataset that contains results from an experiment on plant growth. The yield of a plant is measured by the dried weight of the plant. The experiment recorded yields of plants for a control group and two different treatments. Use a randomization test to analyze the differences in mean yield for the three groups. Start with the exploratory data analysis. What conclusions can be inferred from this sample data?

**10.2 (*flicker* data).** Refer to the `flicker` data in Example 8.1. Use a randomization test to compare the critical flicker frequency among groups with different eye colors. State the null hypothesis and your conclusions carefully.

**10.3 (Waste Run-up data).** Refer to the *Waste Run-up* data in Example 10.1 and Exercise 8.6. Write R code using `replicate` (but without using the function `rand.oneway` in this chapter) to implement a randomization test to determine if mean Waste Run-up differs among plants. Then compare your randomization test result to the result of the function `rand.oneway` in this chapter. The observed $F$ statistics should agree. Explain why the $p$-values could differ.

**10.4 (*airquality* data).** Consider the response variable Ozone in the *airquality* data set in R. Suppose that we are interested in testing the hypothesis that Ozone differs by Month. Do the assumptions for an ANOVA $F$-test appear to hold for this data? Carry out the one-way analysis using appropriate methods for valid inference.

**10.5 (Web hits).** Another way of exploring the pattern of website counts in Example 10.3 is to look for differences between seasons. Weeks 1 through 9 correspond to spring months, weeks 10 through 22 to summer months, and weeks 23 through 35 to fall months. Implement a randomization test to see if website counts differ between seasons.

**10.6 (Web hits, cont.).** In Example 10.3, revise the `rand.correlation` function so that it also displays a probability histogram of the generated rank statistics in `stats`. Identify the observed statistic $r_{obs}$ on the plot using `abline(v = observed.r)`.

**10.7 (Computing rank correlations).** For the website hits data in Example 10.3, we could have used the result of Spearman's rank correlation statistic (`cor` with `method="spearman"`) to handle the calculation of the rank correlation statistics. Revise the function `rand.correlation` in Example 10.3 so that it uses `cor` to compute the rank correlation. For example, if our paired data is in vectors `x` and `y`, then we can obtain the rank correlation by

```
r = cor(x, y, method="spearman")
```

**10.8 (Permutation test for correlation).** Generate two independent random samples of count data using one of the R probability generators. Independent variables are uncorrelated. Write a function to implement a permutation test for correlation applied to the data (not the ranks of the data) and apply it to 50 pairs of your randomly generated data. This exercise is similar to Example 10.3, if we replace the observed web hits and weeks data with two simulated samples.

# Chapter 11
# Simulation Experiments

## 11.1 Introduction

Simulation provides a straightforward way of approximating probabilities. One simulates a particular random experiment a large number of times, and a probability of an outcome is approximated by the relative frequency of the outcome in the repeated experiments. The idea of using chance devices to model random experiments has a long history. During the Second World War, physicists at the Los Alamos Scientific Laboratory used simulation to learn about the characteristics of the travel of neutron through different materials. Since this work was secret, it was necessary to use a code name and the name "Monte Carlo" was chosen by the scientists in homage to the famous Monte Carlo casino. Since that time, the use of simulation experiments to better understand probability patterns is called the Monte Carlo method.

We focus on two particular R functions that simplify the process of programming simulation experiments. The `sample` function will take samples with or without replacement from a set, and the `replicate` function is helpful in repeating a particular simulation experiment. We illustrate the use of these two functions in simulating some famous probability problems.

## 11.2 Simulating a Game of Chance

*Example 11.1 (A coin-tossing game).*
Peter and Paul play a simple game involving repeated tosses of a fair coin. In a given toss, if heads is observed, Peter wins $1 from Paul; otherwise if tails is tossed, Peter gives $1 to Paul. (This example is described in Chapter 1 of Snell [45].) If Peter starts with zero dollars, we are interested in his change in fortune as the game is played for 50 tosses.

## 11.2.1 The sample *function*

We can simulate this game using the R function `sample`. Peter's winning on a particular toss will be $1 or $-$$1 with equal probability. His winnings on 50 repeated tosses can be considered to be a sample of size 50 selected with replacement from the set $\{$1, -$1\}$. This process can be simulated in R by the `sample` command; the arguments indicate we are sampling from the vector $(-1, 1)$, a sample of `size` 50 is taken, and `replace=TRUE` indicates sampling with replacement.

```
> options(width=60)
> sample(c(-1, 1), size=50, replace=TRUE)
 [1] -1 -1 -1  1  1  1 -1  1 -1 -1 -1 -1 -1 -1  1  1  1  1
[19]  1  1 -1 -1 -1  1  1 -1 -1  1  1  1  1  1  1  1  1  1
[37] -1 -1  1  1 -1  1  1  1 -1 -1 -1  1 -1 -1
```

We see that Peter lost the first three tosses, won the next three tosses, and so on.

## 11.2.2 Exploring cumulative winnings

Suppose Peter is interested in his cumulative winnings as he plays this game. We store his individual toss winnings in the variable `win`. The function `cumsum` will compute the cumulative winnings of the individual values and the cumulative values are stored in `cum.win`.

```
> win = sample(c(-1, 1), size=50, replace=TRUE)
> cum.win = cumsum(win)
> cum.win
 [1] -1 -2 -3 -2 -1  0 -1  0 -1 -2 -3 -4 -5 -6 -5 -4 -3 -2
[19] -1  0 -1 -2 -3 -2 -1 -2 -3 -2 -1  0  1  2  3  4  5  6
[37]  5  4  5  6  5  6  7  8  7  6  5  6  5  4
```

In this particular game, we see Peter reaches a negative fortune of $-6$ during the early part of the game and rebounds for a positive fortune of 8 towards the end of the game, finishing with a fortune of 4. Figure 11.1 displays the sequence of cumulative winnings for four simulated games of heads or tails using the following R commands. The `par(mfrow=c(2,2))` command splits the graphics window into subwindows with two rows and two columns so one can display four graphs in one window. After the graphs have been drawn, the `par(mfrow=c(1,1))` command restores the graphics window to the initial state.

```
> par(mfrow=c(2, 2))
> for(j in 1:4){
+   win = sample(c(-1, 1), size=50, replace=TRUE)
+   plot(cumsum(win), type="l" ,ylim=c(-15, 15))
+   abline(h=0)
```

```
+ }
> par(mfrow=c(1, 1))
```

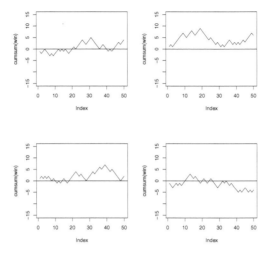

**Fig. 11.1** Graphs of cumulative winnings of Peter in four simulations of the heads or tails game.

The graphs in Figure 11.1 show that there is much variability in Peter's fortune in playing this game. In some games, he tends to break even and his cumulative winnings fall near the horizontal line at zero. In other games, he will be fortunate and the cumulative winnings will be positive for all tosses in the game, and other games he'll have negative winnings for all tosses. After one observes a number of these graphs, one can pose the following questions:.

- What is the chance that Peter will break even after 50 tosses?
- What are likely number of tosses where Peter will be in the lead?
- What will be the value of Peter's best fortune during the game?

Some of these questions (such as the first one) can be found through exact calculation. Other questions such as the second or third are more difficult to find analytically and the Monte Carlo method is helpful for finding approximate answers.

### 11.2.3 R function to implement a Monte Carlo experiment

One can obtain approximate answers to these questions by a Monte Carlo experiment. In this type of experiment, one simulates the random process and computes the statistic or statistics of interest. By repeating the random process many times, one obtains a collection of the statistics. One then uses the collection to approximate probabilities or expectations that answer the questions.

Let's first consider Peter's fortune $F$ at the end of the game. We write a function **peter.paul** that simulates the fortunes for the 50 tosses and computes $F$ that is equal to the sum of the individual fortunes. To make this function more general, we define **n** to be the number of tosses and let the default value of **n** be 50.

```
peter.paul=function(n=50){
  win = sample(c(-1, 1), size=n, replace=TRUE)
  sum(win)
}
```

We simulate a single game by typing

```
> peter.paul()
[1] -6
```

In this game, Peter finished with a fortune of $-6$ dollars. To repeat this for 1000 games, we use the **replicate** function with arguments 1000, the number of games, and the name of the function **peter.paul()** to repeat. The output of **replicate** is the vector of fortunes for the 1000 games that we assign to the variable **F**.

```
> F = replicate(1000, peter.paul())
```

### 11.2.4 Summarizing the Monte Carlo results

Since Peter's fortune is integer-valued, a convenient way of summarizing the collection of values of $F$ using the **table** function.

```
> table(F)
F
-22 -20 -18 -16 -14 -12 -10  -8  -6  -4  -2   0   2   4   6
  2   2   5  10   7  33  37  48  76  94 126 110 109 100  74
  8  10  12  14  16  18  20  22  24
 63  40  27  18  12   3   2   1   1
```

We can display the frequencies of $F$ using the **plot** function applied to the **table** output. (See Figure 11.2.)

```
> plot(table(F))
```

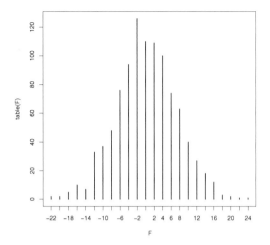

**Fig. 11.2** Line graph of Peter's total fortune in 1000 simulated games of heads or tails.

We see that Peter's fortune has a symmetric distribution centered about zero. Note that no games result in an odd-valued fortune, indicating that is impossible for Peter to finish with a "odd" fortune.

What is the chance that Peter breaks even in the game? We see from the table that Peter finished with a value of $F = 0$ in 110 out of the 1000 simulations, and so the probability of breaking even is approximated by

$$P(F = 0) \approx \frac{110}{1000} = 0.110.$$

Here an exact answer to this probability is available. Peter breaks even if there are exactly 25 heads in the binomial experiment of 50 trials with a probability of success equal to 0.5. Using the R binomial probability function `rbinom`, we obtain

```
> dbinom(25, size=50, prob=0.5)
[1] 0.1122752
```

This exact probability agrees to within two significant digits with the value obtained in our simulation.

## 11.2.5 Modifying the experiment to learn about new statistics

We can add additional lines of code to our function `peter.paul` to compute several statistics of interest in our experiment. To answer our questions, we focus on the final fortune $F$, the number of times Peter is in the lead $L$, and the maximum cumulative winning $M$. In the function, we define the vector of cumulative winnings `cum.win`. Here the output of the function is a vector consisting of the values of $F, L$, and $M$. By naming the components (using, for example, `F=sum(win)`), one will get more attractive output.

```
peter.paul=function(n=50){
  win=sample(c(-1, 1), size=n, replace=TRUE)
  cum.win = cumsum(win)
  c(F=sum(win), L=sum(cum.win > 0), M=max(cum.win))
}
```

We simulate the game once.

```
> peter.paul()
  F   L   M
-10   3   1
```

In this game, Peter's final fortune was $-\$10$, he was in the lead for 3 plays, and his maximum total winning during the game was \$1. To repeat this game 1000 times, we again use `replicate` and store the output in the variable S. Since the output of `peter.paul` is a vector, S will be a matrix of 3 rows and 1000 columns, where the rows correspond to the simulated draws of $F$, $L$, and $M$. We can verify the dimension of the matrix of S using the `dim` function.

```
> S = replicate(1000, peter.paul())
> dim(S)
[1]    3 1000
```

Recall that we were interested in finding likely values of the number of times $L$ that Peter was in the lead. The simulated values of $L$ can be found by accessing the "L"th or second row of S:

```
> times.in.lead = S["L", ]
```

We tabulate the simulated values using the `table` function, and the `prop.table` function will find the corresponding relative frequencies. We `plot` the result to get a line graph. (See Figure 11.3.)

```
> plot(prop.table(table(times.in.lead)))
```

The pattern of this graph for $L$ isn't what most people would expect. Since it is a fair game, one might think it would be most likely for Peter to be ahead in 25 out of the 50 tosses. Instead, the most likely values are the extreme values $L = 0, 1, 50$, and the remaining values of $L$ appear equally likely. So actually it is relatively common for Peter to always be losing or always be winning in this game.

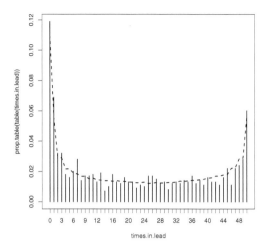

**Fig. 11.3** Line graph of the number of times Peter was in the lead in 1000 simulated games of heads or tails. The overlayed line corresponds to more precise estimates of the probabilities found using a large number of simulations.

In any simulation, there will be an inherent error in any computed probabilities. To see if the variability in the heights of the lines is due to real differences or sampling variability, we repeated the simulation experiment using 100,000 iterations. The dashed line in Figure 11.3 corresponds to the estimated probabilities for this second simulation that are close to the true probabilities. We see that the general impression from our first simulation is correct and the extreme values of $L$ are most likely.

Last, let's consider the distribution of $M$, Peter's maximum winning during the 50 plays. We store the 1000 simulated values of $M$ in the variable `maximum.lead`, tabulate the values using the `table` function, and then graph the values by the `plot` function.

```
> maximum.lead = S["M", ]
> plot(table(maximum.lead))
```

From Figure 11.4, we see that it is most likely for Peter to have a maximum winning of 3 or 4, but values between −1 and 8 are all relatively likely. To compute the approximate probability that Peter will have a maximum winning of 10 or more, we find the number of values of `maximum.lead` that are 10 or greater and divide this sum by the number of simulation iterations 1000.

```
> sum(maximum.lead >= 10) / 1000
[1] 0.149
```

Since this probability is only about 15%, it is relatively rare for Peter to have a maximum winning of 10 or more.

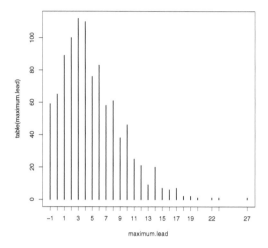

**Fig. 11.4** Line graph of the maximum winning of Peter in 1000 simulated games of heads or tails.

## 11.3 Random Permutations

*Example 11.2 (The hat problem).*

We next consider a famous probability problem. In the old days, a man would wear a top hat to a restaurant and check his hat with a person at the door. Suppose $n$ men check their hats and the hats are returned to the men in a random fashion when they leave the restaurant. How many men will receive their own hats?

### 11.3.1 Using sample to simulate an experiment

We represent the distinct hats of the $n$ men by the integers 1, ..., $n$. We first assume we have $n = 10$ hats and store these integers in the vector **hats**.

```
> n = 10
> hats = 1:n
```

Randomly mixing the hats corresponds to a random permutation of the integers that we accomplish using the simple function **sample** – the vector of mixed-up integers is stored in the variable **mixed.hats**. (Note that when **sample** is used with only a single vector argument, the result is a random permutation of the elements of the vector.)

```
> mixed.hats = sample(hats)
```

We display both vectors below and note that in this particular simulation, the 4th and 9th hats were returned to the correct owners.

```
> hats
 [1]  1  2  3  4  5  6  7  8  9 10
> mixed.hats
 [1]  7  1  6  4  3  5  8 10  9  2
```

## 11.3.2 Comparing two permutations of a sample

We are interested in computing the number of correct hats returned. If we use the logical expression

```
> hats == mixed.hats
 [1] FALSE FALSE FALSE  TRUE FALSE FALSE FALSE FALSE  TRUE FALSE
```

we obtain a vector of logical values, where **TRUE** corresponds to a match and **FALSE** corresponds to a mismatch. If we sum these logical values (using the function sum), **TRUE** is replaced by a 1, **FALSE** is replaced by a 0, and the result is the number of correct matches that is stored in a variable `correct`. Displaying this variable, it correctly tells us that two men received their own hats.

```
> correct = sum(hats == mixed.hats)
> correct
[1] 2
```

$\mathbf{R_x}$ *11.1 What happens if one incorrectly types this line using a single equals sign instead of a double equals sign?*

```
> correct = sum(hats = mixed.hats)
> correct
[1] 55
```

*Surprisingly, there is no error reported and an "answer" of 55 is displayed. What is going on? The argument hats=mixed.hat actually only labels the individual values of the vector, and the sum functions sums the integers from 1 to 10 which is 55. This is clearly not we wanted to do!*

## 11.3.3 Writing a function to perform simulation

Using the above R commands, we can write a function that performs this mixed-up hats simulation. We call this function `scramble.hats` and it has a single argument $n$, the number of hats. The function returns the number of correct matches.

```
scramble.hats = function(n){
  hats = 1:n
  mixed.hats = sample(n)
  sum(hats == mixed.hats)
}
```

We test this function once using $n = 30$ hats; in this particular simulation, we see that no correct hats were returned.

```
> scramble.hats(30)
[1] 0
```

## 11.3.4 Repeating the simulation

Let $X_n$ denote the random variable defined as the number of correct matches for a given number of hats $n$. To obtain the probability distribution of $X_n$, we simulate this experiment (coded in the function scramble.hats) a large number of times and tabulate the simulated values of $X_n$. Repeated random experiments are easily performed in R by the function replicate which has two arguments, the number of simulations, and the expression or function to be repeated. If we wish to repeat our hats experiment 1000 times with $n = 10$ hats, we type

```
> matches = replicate(1000, scramble.hats(10))
```

The vector of number of matches in all 1000 experiments is stored in the variable matches.

To learn about the probability distribution of $X_n$, we summarize the simulated draws of $X_n$ in the vector matches. By tabulating the values (using the table function), we obtain a frequency table of the number of matches that occurred.

```
> table(matches)
matches
  0   1   2   3   4   5   6
393 339 179  66  19   3   1
```

By dividing the table frequencies by 1000, we obtain the corresponding relative frequencies.

```
> table(matches) / 1000
matches
    0     1     2     3     4     5     6
0.393 0.339 0.179 0.066 0.019 0.003 0.001
```

We see that it is most likely (with estimated probability 0.393) to obtain no matches and it is relatively uncommon to get 3 or more matches. To obtain the (approximate) mean or expected value of $X_n$, we simply find the sample mean of the simulated draws:

```
> mean(matches)
[1] 0.992
```

On average, we expect that 0.992 of the 10 men will receive his correct hat. It can be shown that the theoretical mean is indeed $E(X_n) = 1$ (for all $n$), so this simulation has accurately estimated the mean number of correct matches.

Suppose we are interested in seeing how the probability of no correct matches, $P(X_n = 0)$, changes as a function of the number of men $n$. We write a new function prop.no.matches that has a single argument $n$; the function simulates this experiment (with $n$ men) 10,000 times and finds the proportion of experiments where no men received the correct hats.

```
prop.no.matches = function(n){
  matches = replicate(10000, mixed.hats(n))
 sum(matches == 0) / 10000
}
```

We test this function for $n = 20$ hats:

```
> prop.no.matches(20)
[1] 0.3712
```

In this experiment, the chance that no men received their correct hats was 0.3712. The function sapply is a convenient way of applying this function for a vector of values of $n$. To compute this probability for values of $n$ from 2 to 20, we use sapply with arguments the vector of values and the function name of interest. In Figure 11.5 we display the estimated probabilities as a function of $n$ and add a horizontal line at the value 0.37.

```
> many.probs = sapply(2:20, prob.no.matches)
> plot(2:20, many.probs,
+   xlab="Number of Men", ylab="Prob(no matches)")
> abline(h=0.37)
```

In Figure 11.5, the probability of no matches is seen to be large for $n = 2$, small for $n = 3$, and the probability appears to stabilize about the value of 0.37 for large $n$.

## 11.4 The Collector's Problem

*Example 11.3 (Collecting baseball cards).*
    One of the authors collected baseball cards as a youth. The baseball card company Topps would produce a collection of cards consisting of cards for each professional player together with some special cards. One would purchase cards in small packs consisting of ten cards and a stick of bubble gum. (The author didn't chew gum and the stick of gum was promptly thrown away.) One objective was to obtain a complete set of cards. Unfortunately, it was common to obtain duplicate cards and it seemed that one had to purchase a large number of cards to obtain a complete set.

## 11.4.1 Simulating experiment using the sample function

One can illustrate this card collecting activity through a simulation. Suppose we represent the complete set by a variable **cards**, a character vector consisting of the names of ten great baseball hitters.

```
> cards = c("Mantle", "Aaron", "Gehrig", "Ruth", "Schmidt",
+      "Mays", "Cobb", "DiMaggio", "Williams", "Foxx")
```

Assume that each purchased card is equally likely to be one of the values in **cards**. Assuming independent draws, a purchase of 20 cards can be considered a random sample taken with replacement from this vector. This is accomplished by the **sample** function; the arguments are the "population" vector **cards**, the size of the sample (20), and **replace=TRUE**, indicating that the sample is to be taken with replacement. We display the 20 cards that are purchased.

```
> samp.cards = sample(cards, size=20, replace=TRUE)
> samp.cards
 [1] "Foxx"     "DiMaggio" "Aaron"    "Schmidt"  "DiMaggio"
 [6] "Mays"     "Mays"     "Cobb"     "Mantle"   "Foxx"
[11] "Schmidt"  "Mays"     "DiMaggio" "Foxx"     "Gehrig"
[16] "Schmidt"  "Mantle"   "DiMaggio" "Mays"     "Foxx"
```

We note that we obtained some duplicates – for example, we got four DiMaggio's and four Foxx's. Did we purchase a complete set? One way to

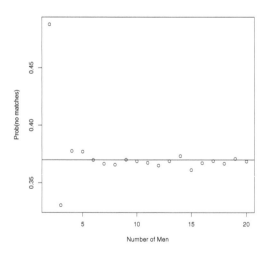

**Fig. 11.5** Monte Carlo estimates of the probability of no matching hats as a function of the number of hats $n$.

check is to apply the `unique` function that finds the unique values in the vector, and then use the `length` function to compute the length of this vector.

```
> unique(samp.cards)
[1] "Foxx"     "DiMaggio" "Aaron"     "Schmidt"  "Mays"
[6] "Cobb"     "Mantle"   "Gehrig"
> length(unique(samp.cards))
[1] 8
```

Here we did not purchase a complete set, since there were only eight unique cards out of a possible ten.

## 11.4.2 Writing a function to perform the simulation

In general, suppose that a complete set consists of $n$ cards, $m$ cards are purchased, and we are interested in the probability that a complete set is obtained. Our author was actively buying baseball cards in 1964 to complete a Topps set consisting of 586 cards. If he purchased 3000 cards, what is the chance he would obtain a complete set? We write a function `collector` that simulates this process. The complete set is represented by the vector consisting of the integers from 1 to $n$. We use the `sample` function to take the sample of $m$ cards. The `ifelse` function is used to check if we have purchased a complete set; if the number of unique cards in our sample is equal to $n$, the function returns "yes," otherwise the function returns "no."

```
collector=function(n,m){
  samp.cards = sample(1:n, size=m, replace=TRUE)
  ifelse(length(unique(samp.cards)) == n, "yes", "no")
}
```

To simulate this experiment once, this function is applied with the arguments $n = 586$ and $m = 3000$:

```
> collector(586, 3000)
[1] "no"
```

In this experiment, the outcome is "no" – the author did not get a complete set.

To see if the author was unlucky, we use the `replicate` function to repeat this process for 100 experiments. We use the `table` function to summarize the outcomes.

```
> table(replicate(100, collector(586, 3000)))
 no yes
 97   3
```

In these 100 experiments, a complete set was found three times and the estimated probability of purchasing a complete set is $3/100 = 0.03$.

## 11.4.3 Buying an optimal number of cards

Let's make this problem a little more interesting. Suppose our card collector is reluctant to purchase an arbitrarily large number of cards to be sure to get a complete set. Suppose that the cards from the bubble gum packs cost a nickel each, but he has the opportunity to purchase individual cards from a dealer at a cost of 25 cents each. Suppose our collector plans on purchasing a fixed number of cards with the intention of purchasing "missing" cards from the dealer to complete the set. On average, what is the cost of this plan? Is there an optimal number of cards purchased that will minimize the expected cost?

We write a function `collect2` to simulate this card buying process. The function has a single argument `n.purchased`, the number of cards purchased. We assume the cost of a "random" card is `cost.rcard = 0.05` and the cost of a card from the dealer is `cost.ncard = 0.25`. As in the earlier simulation, we sample `n.purchased` cards with replacement from the complete set represented by the integers from 1 to 586. We compute the number of (random) unique cards we collect `n.cards` and the number of cards we haven't collected `n.missed`. Assuming we purchased the remaining cards from the dealer, the random total cost will be

$$COST = \texttt{cost.rcard} \times \texttt{n.purchased} + \texttt{cost.ncard} \times \texttt{n.missed}.$$

```
collect2 = function(n.purchased){
  cost.rcard = 0.05
  cost.ncard = 0.25
  samp.cards = sample(1:586, size=n.purchased, replace=TRUE)
  n.cards = length(unique(samp.cards))
  n.missed = 586 - n.cards
  n.purchased * cost.rcard + n.missed * cost.ncard
}
```

Let's illustrate using the function `collect2` in several ways. Suppose one decides to purchase 800 cards. What is the expected cost? We can repeat the random process of purchasing cards using the `replicate` function – below this process is repeated 500 times and the vector of total costs is stored in the variable `costs`. The distribution of random costs is summarized by the `summary` function.

```
> costs = replicate(500, collect2(800))
> summary(costs)
   Min. 1st Qu.  Median    Mean 3rd Qu.    Max.
   69.5    76.0    77.2    77.3    78.8    83.8
```

We see the cost of buying 800 cards has a 50% chance of falling between $76.00 and $78.80. The expected cost of buying 800 cards is estimated by the sample mean of $77.30.

Since we are primarily interested in the expected cost, we write a new function `expected.cost` that takes 100 samples of cards, each of size `n.purchased` and computes the average total cost. Note that the function `replicate` is used to do the repeated simulations and the `mean` function finds the mean of the vector of total costs.

```
expected.cost = function(n.purchased)
  mean(replicate(100, collect2(n.purchased)))
```

Since we are interested in seeing how the total cost varies as a function of the number purchased, we define a vector N containing plausible numbers of cards to purchase from 500 to 1500. Using the `sapply` function, we apply the function `expected.cost` to each of the values in N and the resulting average costs are placed in the vector ECOST. A plot of the expected cost against the number of cards purchased is constructed in Figure 11.6. (The `grid` function adds an overlaid grid to the display to help locate the minimum value of the function.)

```
> N=500:1500
> ECOST = sapply(N, expected.cost)
> plot(N, ECOST, xlab="Cards Purchased",
+   ylab="Expected Cost in Dollars")
> grid(col="black")
```

From the graph, the expected cost appears to be minimized at approximately 950 cards. So the optimal number of cards to purchase is about 950 and the expected cost using this strategy is about $76.50. If we decide instead to purchase 1200 cards, the expected cost would be about $79.

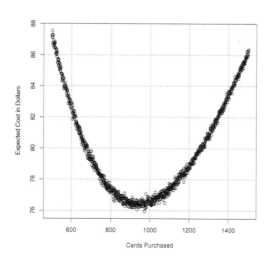

**Fig. 11.6** Expected total cost of completing a baseball card collection as a function of the number of "random" cards purchased where the expected cost is computed by a Monte Carlo experiment.

## 11.5 Patterns of Dependence in a Sequence

*Example 11.4 (Streaky hitting in baseball).*

Sports fans are fascinated with athletes who exhibit patterns of extreme performance. Athletes go through periods of time where they are performing unusually well or unusually poorly, and scorekeepers keep records of these extreme patterns. In the game of baseball, a batter wishes to get a base hit during a game, and record keepers keep track of the number of consecutive games where a player gets a base hit – this is called a hitting streak. During the 2006 baseball season, Chase Utley of the Philadelphia Phillies had a hitting streak of 35 games which is one of the best hitting streaks in baseball history.

But how "significant" is Utley's 35-game hitting streak? Chase Utley was a pretty good hitter during this season so one would expect him to have long hitting streaks. In addition, long runs of heads or tails can be observed in repeated flips of a fair coin. So perhaps Utley's 35-game hitting streak was not as significant as it appears. We will use a Monte Carlo simulation to better understand the patterns of streaky hitting in baseball.

### 11.5.1 Writing a function to compute streaks

Suppose we represent a baseball player's hitting success or failures in a series of games by a sequence of 0's and 1's, where 1 represents a base hit in that particular game. Consider a hypothetical hitter who plays in 15 games and observes the following sequence of games with hits and no hits that we store in the vector y.

```
> y = c(1, 1, 1, 1, 1, 1, 1, 0, 0, 1, 0, 1, 1, 1, 1)
```

We are interested in recording the lengths of all hitting streaks in this sequence. First, we add a 0 to the beginning and end of the sequence to bracket the strengths and `where` is a logical vector which is `TRUE` or `FALSE` if a 0 or 1 is observed. The vector `loc.zeros` records the locations in the sequence where 0's are observed.

```
> where = (c(0, y, 0) == 0)
> n = length(y)
> loc.zeros = (0:(n+1))[where]
> loc.zeros
[1]  0  8  9 11 16
```

We see that this particular hitter did not get a hit in games 0, 8, 9, 11, and 16. The lengths of the hitting streaks are found by computing the differences in this sequence, subtracting one, and removing the values that are equal to zero.

```
> streak.lengths = diff(loc.zeros) - 1
> streak.lengths = streak.lengths[streak.lengths > 0]
> streak.lengths
[1] 7 1 4
```

This player had hitting streaks of length 7, 1, and 4. We can find the longest hitting streak by taking the `max` of this vector:

```
> max(streak.length)
[1] 7
```

We create a function `longest.streak` that uses the above commands to compute the longest streak of 1's in a binary vector `y`.

```
longest.streak=function(y){
  where = c(0, y, 0) == 0
  n = length(y)
  loc.zeros = (0:(n+1))[where]
  streak.lengths = diff(loc.zeros) - 1
  streak.lengths = streak.lengths[streak.lengths > 0]
  max(streak.lengths)
}
```

To check this function, we read in the file "utley2006.txt" that contains the game by game hitting data for Chase Utley for the 2006 baseball season; this data is stored in the data frame `dat`. The variable H gives the number of hits in each game. We create the binary vector by first finding a logical vector indicating if H is positive and then using the `as.numeric` function to convert TRUEs and FALSEs to 1's and 0's. We apply the function `longest.streak` to the sequence `utley` and we confirm that Utley had a 35-game hitting streak this season.

```
> dat = read.table("utley2006.txt", header=TRUE, sep="\t")
> utley = as.numeric(dat$H > 0)
> longest.streak(utley)
[1] 35
```

## 11.5.2 Writing a function to simulate hitting data

Now we are ready to use a Monte Carlo simulation to understand the "significance" of Utley's hitting streak. In this particular season, Utley played in 160 games and hit in 116 of these games. Suppose that these 116 "hit games" were randomly distributed throughout the sequence of 160 games. What would be the length of the longest streak of this random sequence? This is easily answered by a simulation that we code in a function `random.streak` with the binary sequence `y` as an argument. This function first randomly permutes the binary sequence using the `sample` function; the random sequence is stored in the vector `mixed.up.y`. Then the longest streak of 1's in the random sequence is computed using the function `longest.streak`.

```
random.streak=function(y){
  mixed.up.y = sample(y)
  longest.streak(mixed.up.y)
}
```

We use the `replicate` function to repeat this simulation in `random.streak` 10,000 times using Utley's hitting sequence. We store the longest hitting streaks in these 10,000 simulations in the vector L. We tabulate the values of L, graph the tabulated values using the `plot` function, and overlay Utley's longest hitting streak of 35 by a vertical line.

```
> L = replicate(100000, random.streak(utley))
> plot(table(L))
> abline(v=35, lwd=3)
> text(38, 10000, "Utley")
```

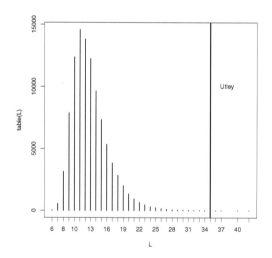

**Fig. 11.7** Graph of the length of the longest hitting streaks of sequences based on random permutations of Utley's hitting sequence. Utley's actual longest hitting streak of 35 games is displayed.

What have we learned? If the games where Utley received a hit were randomly distributed through the 160 games of the season, then from Figure 11.7 we see that longest hitting streaks from 8 to 18 are quite common and a hitting streak of 11 games is most likely. Utley's 35 game hitting streak is in the far right tail of this distribution, indicating that his hitting streak would be very rare among random sequences. One can measure "how extreme" by estimating the probability that a random sequence would have a hitting streak of length 35 or greater. In our particular simulation, one observed 7 hitting streaks of length 35 or greater so the estimated probability is

$$P(\text{hitting streak} \geq 35) \approx \frac{7}{10000} = 0.0007.$$

For a hitter of Utley's ability, the estimated chance of observing a hitting streak at least 35 games is only 0.0007. This conclusion should be made with some caution, since we chose Utley's hitting record since he had a long hitting streak. The chance that *some* hitter during the 2006 baseball would have a hitting streak of 35 or more games is a larger probability, since there are more outcomes involved in the computation.

## Exercises

**11.1 (Playing Roulette).** Suppose one plays roulette repeatedly at a casino. In a single play, one bets $5 on "red"; the player wins $5 with probability 18/38 and loses $5 with probability 20/38. If the roulette game (with the same bet) is played 20 times, then the individual play winnings can be viewed as a sample of size 20 selected with replacement from the vector (5, −5), where the respective probabilities are given in the vector (18/38, 20/38). These play winnings can be simulated using the function `sample` with the `prob` vector that gives the sampling probabilities.

```
sample(c(5, -5), size=20, replace=TRUE,
  prob=c(18 / 38, 20 / 38))
```

a. Write a short function to compute the sum of the winnings from 20 plays at the roulette wheel. Use the `replicate` function to repeat this "20 play simulation" 100 times. Find the approximate probability that the total winning is positive.
b. The number of winning plays is a binomial random variable with 20 trials where the probability of success is 18/38. Using the `dbinom` function, find the exact probability that your total winning is positive and check that the approximate answer in part (a) is close to the exact probability.
c. Suppose you keep track of your cumulative winning during the game and record the number of plays $P$ where your cumulative winning is positive. If the individual play winnings are stored in the vector `winnings`, the expression `cumsum(winnings)` computes the cumulative winnings, and the expression `sum(cumsum(winnings)>0)` computes a value of $P$. Adjust your function from part (a) to compute the value of $P$. Simulate the process 500 times and construct a frequency table of the outcomes. Graph the outcomes and discuss which values of $P$ are likely to occur.

**11.2 (Checking hats).** Suppose that men in the old days wore only two types of hats, say black and grey, and hats of a particular type are indistinguishable. Assume 20 men with hats visit the restaurant and half of the men are wearing each type of hat. The hats are randomly mixed, and we are interested in the number of men who leave the restaurant with the correct hat.

a. Modify the function `scramble.hats` to compute the number of correct matches in this setting. (The only change is the definition of the vector `hats` – if one represents a black hat and a grey hat by 1 and 2, respectively, then `hats` consists of ten 1's and ten 2's.)
b. Using the function `replicate`, repeat this simulation for 1000 trials. Store the number of matches for the 1000 experiments in the vector `matches`.
c. From the simulated values, approximate the probability that 10 or more men receive the correct hats. Also, find the expected number of correct matches.

**11.3 (Birthday problem).** Suppose you have a class of $n$ students and you're interested in the probability that at least two students in the class share the same birthday. If one assumes that each birthday is equally likely from the set $\{1, 2, ..., 365\}$, then collecting birthdays can be viewed as a sample of size $n$ chosen with replacement from the set.

a. Write a function with argument $n$ that samples $n$ birthdays and computes the number of unique birthdays observed.
b. Using the `replicate` function, repeat this simulation 1000 times for the special case of $n = 30$ students.
c. From the output, approximate the probability that there is at least one matching birthday among the 30 birthdays.
d. It can be shown that the exact probability of a birthday match is given by

$$P(match) = 1 - \frac{365 P_{30}}{365^{30}},$$

where $_N P_k$ is the number of ways of arranging a sample of $k$ of $N$ distinct items. Compare the approximate probability in part (d) with the exact probability.
e. The R function `pbirthday`, with argument $n$, computes the (approximate) probability of at least one match among $n$ birthdays. Use this function to check your answer in part (d).

**11.4 (Streakiness).** In the chapter, the focus was on the length of the longest streak in a binary sequence. Another way to measure streakiness in a sequence is the number of switches from 0 to 1 or from 1 for 0. For example, if the binary sequence is given by

```
0  1  0  0  0  1  0  0  1
```

there are three switches from 0 to 1 and two switches from 1 to 0 and so the total number of switches is equal to five. If `y` is a vector containing the binary sequence, then the R expression

```
sum(abs(diff(y)))
```

will compute the number of switches.

a. Construct a function `switches` that computes the number of switches for a binary vector $y$. Test this function by finding the number of switches in Chase Utley's game hitting sequence for the 2006 season.
b. By making a slight change to the function `random.streak`, construct a function that computes the number of switches for a random permutation of the 1's and 0's in the vector $y$.
c. Use the `replicate` function to repeat the random permutation in part (b) for 10,000 simulations. Construct a histogram of the number of switches for these 10,000 random sequences. Is the number of switches in Utley's sequence consistent with the values generated from these random sequences? Using the number of switches statistic, did Utley display unusually streaky behavior during this season?

**11.5 (Collecting state quarters).** In 1999, the United States launched the 50 State Quarters program where each of the 50 states was honored with a special quarter. Suppose you purchase 100 "state" quarters where each quarter is equally likely to feature one of the 50 states.

a. Write a function using the `sample` function to simulate the purchase of 100 quarters and record the number of unique quarters that are purchased.
b. Using the `replicate` function, repeat this process for 1000 purchases. Construct a table of the number of unique quarters you obtain in these 1000 simulations. Use this table to estimate the probability that you obtain at least 45 unique quarters.
c. Use the output from part (b) to find the expected number of unique quarters.
d. Suppose you are able to complete your quarter set by purchasing state quarters from a coin shop for $2 for each quarter. Revise your function to compute the total (random) cost of completing the quarter set. Using the `replicate` function, repeat the quarter-purchasing process 1000 times and compute the expected cost of completing your set.

**11.6 (How many students are excluded?).** In [35], Frederick Mosteller mentioned the following interesting probability question that can be addressed by a simulation experiment. Suppose there is a class of 10 students and each student independently chooses two other class members at random. How many students will not be chosen by any other students?

Here is an outline how one can simulate this experiment, following Mosteller's suggestions:

- One represents the students by a vector consisting of the integers 1 through 10.

  ```
  students = 1:10
  ```

- One represents the outcomes of choosing students by a matrix m of 10 rows and 10 columns consisting of zeros and ones. The $(i, j)$ entry in the matrix is equal to one if student $i$ chooses student $j$. One can initially define the matrix with all zeros by the R code.

```
m = matrix(0, nrow=10, ncol=10)
```

- Student $j$ will choose two other students from the vector **students** excluding the value $j$. This can be done by the **sample** function – the students who are sampled are stored in the vector **s**.

```
s = sample(students[-j], size=2)
```

- In the matrix **m**, one records the selected students by assigning the columns labeled **s** in the $j$th row with the value 1.

```
m[j, s] = c(1, 1)
```

- If the above procedure is repeated for each of the 10 students, then one can represent all selections by the matrix **m**. Here is the matrix for one simulation.

```
        [,1] [,2] [,3] [,4] [,5] [,6] [,7] [,8] [,9] [,10]
 [1,]    0    1    0    0    1    0    0    0    0    0
 [2,]    0    0    0    1    0    0    1    0    0    0
 [3,]    0    0    0    0    0    0    0    1    0    1
 [4,]    1    0    0    0    1    0    0    0    0    0
 [5,]    0    1    0    0    0    1    0    0    0    0
 [6,]    0    0    0    1    1    0    0    0    0    0
 [7,]    0    1    0    0    0    1    0    0    0    0
 [8,]    0    0    1    0    0    1    0    0    0    0
 [9,]    0    1    0    0    0    0    1    0    0    0
[10,]    1    0    0    0    1    0    0    0    0    0
```

By looking at the sum of the matrix over rows, one can see how many students were not chosen by any students. (In this example, one student was not chosen.)

a. Write a function **mosteller** that will run a single simulation of this experiment.
b. Using the **replicate** function, repeat this simulation for 100 iterations.
c. Use the **table** function to tabulate the number of students not chosen over the 100 experiments.
d. What is the most likely number of students not chosen? What is the approximate probability of this outcome?

# Chapter 12
# Bayesian Modeling

## 12.1 Introduction

There are two general approaches in statistical inference. In the inference and regression chapters, we have discussed the familiar frequentist inferential methods such as the t confidence interval, the chi-square test, and the ANOVA test of equality of means. These are called *frequentist* methods since one evaluates the goodness of these methods by their average performance in repeated sampling. For example a 90% confidence interval has the property that the random interval will cover the unknown parameter 90% of the time in repeated sampling. This chapter introduces the second general approach to inference, the Bayesian method.

Bolstad [4] and Hoff [23] present respectively elementary and intermediate level introductions to Bayesian thinking. Albert [1] illustrates the use of R to perform Bayesian computations.

*Example 12.1 (Deciding authorship).*

This chapter discusses the basic components of a Bayesian analysis using data from a famous Bayesian analysis from Mosteller and Wallace [36] about literary style. The *Federalist Papers* were a series of essays, written in 1787-1788, to persuade the New York state citizens to ratify the U.S. Constitution. Of the essays, it is known that Alexander Hamilton was the sole author of 51, James Madison was the sole author of 14, and the two collaborated on another three. The authorship of the remaining 12 papers has been disputed, and the main problem addressed by Mosteller and Wallace was to determine the author of the disputed papers.

## 12.2 Learning about a Poisson Rate

A first step in the authorship study is to examine the frequency of word use of the authors Alexander Hamilton and James Madison. One useful class of words in this study are so-called *function* words – these are words such as *a*, *an*, *by*, *to*, and *than* that are used to connect or amplify nouns and verbs.

Mosteller and Wallace divided a large amount of the writings of James Madison into blocks of 1000 words, and the number of occurrences of the word *from* was observed in each block. Table 12.1 summarizes the frequency of use of the word *from* from 262 blocks of writing. From the table, we see that there were 90 blocks where the word did not appear, 93 blocks where the word appeared exactly one time, 42 blocks where the word appeared two times, and so on.

**Table 12.1** Frequency table of the occurrences of the word *from* in 262 blocks of text from the writings of James Madison.

| Occurrences | | | | | | |
|---|---|---|---|---|---|---|
| 0 | 1 | 2 | 3 | 4 | 5 | 6 |
| observed 90 | 93 | 42 | 17 | 8 | 9 | 3 |

Suppose $y$ represents the number of occurrences of the word *from* in a randomly selected block of text from Madison's writing. A popular probability model for $y$ is the Poisson distribution with rate $\lambda$ with probability function given by

$$f(y|\lambda) = \frac{\exp(-\lambda)\lambda^y}{y!}, \qquad y = 0, 1, 2, \ldots$$

If this Poisson model is a suitable fit to these data, then one can measure Madison's tendency to use this particular word by the rate parameter $\lambda$. Since $\lambda$ is the mean of the Poisson distribution, $\lambda$ represents the average number of occurrences of this particular word in many blocks of Madison writings.

## 12.3 A Prior Density

In the Bayesian approach, one uses the language of subjective probability to represent uncertainty about unknown parameters. One uses subjective probability to express one's degree of belief about the truth of a "one-time event" such as the event that one will get married in five years or that man will visit the planet Mars in the next 20 years. Subjective probability is personal – an instructor's belief that a student will get an A in a class will likely be different than the student's belief about the same event. Also, a person's subjective probability of an event can change, especially when she obtains new information. In fact, Bayes' rule gives us a formal rule for determining how one's probabilities will change after observing new information.

Here the unknown parameter is $\lambda$ that represents the mean number of occurrences of the word *from* in a block of writing. Since $\lambda$ is unknown, Bayesians regard $\lambda$ as a random quantity and one's knowledge about this parameter is represented using a continuous probability density $g(\lambda)$. This is called a *prior* density since it represents one's beliefs about the parameter prior or before any data is collected.

How does one find the prior density? In this example, the researcher is the person doing the Bayesian inference and she has some beliefs about the location of this parameter that are represented by a prior density. The researcher has analyzed patterns of word use for authors who lived in the same era as James Madison and she has some knowledge about possible values for $\lambda$. Specifically, she believes that Madison's mean rate of *from*'s is equally likely to be smaller or larger than 1.0 and she is confident, with probability 0.9, that the mean rate is smaller than 1.5. Using symbols, she believes that

$$P(\lambda < 1.0) = 0.5, \ P(\lambda < 1.5) = 0.9.$$

After the researcher states her prior beliefs, next she wishes to construct a prior density $g(\lambda)$ that matches this information. There are many possible choices for this density. The gamma family is a convenient and well-known probability density for a continuous positive variable; the gamma density with shape $a$ and rate $b$ is defined by

$$g(\lambda) = \frac{b^a \lambda^{a-1} \exp(-b\lambda)}{\Gamma(a)}, \lambda > 0.$$

After some trial and error, she finds that the particular gamma density with shape $a = 8.6$ and rate $b = 8.3$

$$g(\lambda) = \frac{8.3^{8.6} \lambda^{8.6-1} \exp(-8.3\lambda)}{\Gamma(8.6)}, \lambda > 0,$$

has, approximately, a median of 1.0 and a 90th percentile of 1.5, and so this density is a reasonable approximation to the researcher's beliefs about Madison's average use of the word *from*.

## 12.4 Information Contained in the Data: the Likelihood

There are two sources of information about the rate parameter $\lambda$ – the prior information modeled by the density $g(\lambda)$ and the data. One represents the information contained in the data by the *likelihood* function.

Recall that we observe the number of occurrences $y$ of the specific word in many blocks of text. Assuming independence between observations on different blocks, the probability of observing a collection of $y$ values is given by

$$Prob(\text{data}|\lambda) = \prod_y f(y|\lambda),$$

where $f(y|\lambda)$ is the Poisson density with rate $\lambda$. Substituting the Poisson formula, we obtain

$$Prob(\text{data}|\lambda) = \prod_y f(y|\lambda)$$

$$= \prod_y \frac{\exp(-\lambda)\lambda^y}{y!}$$

$$= C\exp(-n\lambda)\lambda^s,$$

where $n$ is the number of blocks, $s$ is the total number of words counted, and $C$ is a positive constant. In the data displayed in Table 12.1, we see that $n = 262$ and $s = 323$, and this probability is given by

$$Prob(\text{data}|\lambda) = C\exp(-262\lambda)\lambda^{323}.$$

The *likelihood* is the probability of the data outcome $Prob(\text{data}|\lambda)$ viewed as a function of the parameter $\lambda$. We represent the likelihood by the symbol $L$ – here

$$L(\lambda) = \exp(-262\lambda)\lambda^{323}.$$

The likelihood function tells us what values of the parameter $\lambda$ are likely and not likely given the observed data set. This function is helpful in comparing the relative plausibility of different parameter values (given the data), and multiplying the likelihood by any positive constant won't change this relative comparison.

It is informative to plot the likelihood as a function of the parameter $\lambda$. Using the `curve` function, the likelihood $L(\lambda)$ is displayed for values of $\lambda$ between 0.9 and 1.6. (See Figure 12.1.)

```
> curve(exp(-262 * x) * x^323, from=0.9, to=1.6,
+    xlab="Lambda", ylab="Likelihood")
```

We see from the figure that the parameter value $\lambda = 1.2$ is close to the most likely parameter value given this particular data, and, in contrast, parameter values of $\lambda = 1.0$ and $\lambda = 1.5$ are not likely values, which means that these parameter values are not consistent with the observed data.

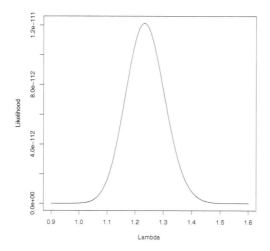

**Fig. 12.1** Likelihood function for the rate of using the word *from* in Madison's writings.

## 12.5 The Posterior and Inferences

### *12.5.1 Computation of the posterior*

The prior density $g(\lambda)$ reflect the expert's opinion about the rate parameter before any data is collected. After the data has been observed, one's opinions about the parameter will change – the new or updated density is called the *posterior* density. Using Bayes' rule, the posterior density is found by the simple formula

$$\text{posterior density} \propto \text{likelihood} \times \text{prior density}.$$

In our example, the expert's prior density was given (ignoring constants) by

$$g(\lambda) \propto \lambda^{8.6-1} \exp(-8.3\lambda),$$

and the likelihood function is given by

$$L(\lambda) \propto \exp(-262\lambda)\lambda^{323}.$$

So by multiplying the prior density and the likelihood, the posterior density is given by

$$g(\lambda|\text{data}) \propto \left[\lambda^{8.6-1}\exp(-8.3\lambda)\right] \times \left[\exp(-262\lambda)\lambda^{323}\right]$$
$$= \lambda^{331.6-1}\exp(-270.3\lambda)$$

The posterior density reflects the updated or current beliefs about the rate parameter $\lambda$ after observing the data. All inferences about the parameter, including point and interval estimates, are based on different summaries of this density. We illustrate Bayesian inference computations using several methods. The first computational method recognizes that the posterior density is a familiar functional form and one can summarize the density using special R functions for this functional form. The second computational method is based on simulation and motivates a general strategy for summarizing a wide range of posterior densities.

### 12.5.2 Exact summarization of the posterior

The gamma density has the functional form

$$f(y) \propto \lambda^{a-1}\exp(-b\lambda),\ y > 0,$$

where $a$ and $b$ are respectively the shape and rate parameters. Comparing this functional form with our posterior density in our example, we recognize that the posterior for $\lambda$ is a gamma density with shape $a = 331.6$ and rate $b = 270.3$. Using the R probability functions for the gamma distribution, it is straightforward to summarize the posterior density.

One communicates a Bayesian inference by simply plotting the posterior density of the parameter. A graph of the posterior density of $\lambda$ can be constructed using the **curve** function and the gamma density function **dgamma**. By trial and error, the **to=1.0** and **from=1.5** arguments in **curve** produce a good display of the posterior density. We superimposed the likelihood function on this graph using the **curve** function with the **add=TRUE** option. (See Figure 12.2.) Note that the likelihood and posterior density are very similar, indicating that the posterior is controlled mainly by the information in the data and the prior information has a minimal effect on the inference.

```
> curve(dgamma(x, shape=331.6, rate= 270.3), from=0.9, to=1.6)
> curve(dgamma(x, shape=324, rate=262), add=TRUE, lty=2, lwd=2)
> legend("topright", c("Posterior", "Likelihood"), lty=c(1, 2),
+   lwd=c(1, 2))
```

A Bayesian point estimate of a parameter is a summary of the posterior density. One useful point estimate of the rate parameter $\lambda$ is the posterior median – this is the value $\lambda_M$ such that

$$P(\lambda < \lambda_M) = 0.5.$$

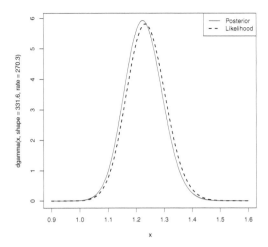

**Fig. 12.2** Posterior and the likelihood function for the rate of using the word *from* in Madison's writings.

We find the posterior median using the gamma quantile function qgamma with inputs 0.5 (the probability) and the shape and rate parameters.

```
> qgamma(0.5, shape=331.6, rate=270.3)
[1] 1.225552
```

Our opinion after observing the data is that Madison's rate is equally likely to be smaller or larger than 1.226.

A Bayesian interval estimate for the rate parameter is an interval $(\lambda_{LO}, \lambda_{HI})$ that contains $\lambda$ with a specific probability $\gamma$:

$$P(\lambda_{LO} < \lambda < \lambda_{HI}) = \gamma.$$

Suppose we want a 90% Bayesian interval where $\gamma = 0.90$. An "equal-tails" interval is found by computing the 5th and 95th percentiles of the gamma posterior density using a second application of the qgamma function:

```
> qgamma(c(0.05, 0.95), shape=331.6, rate=270.3)
[1] 1.118115 1.339661
```

The posterior probability that Madison's rate parameter $\lambda$ is in the interval (1.118, 1.340) is 0.90.

### 12.5.3 Summarizing a posterior by simulation

Simulation provides an attractive general method of summarizing a posterior probability distribution. One simulates a large number of draws of the parameter from the posterior probability distribution and one performs Bayesian inference by finding appropriate summaries of the simulated sample.

In our example, the posterior density is a gamma(331.6, 270.3) density, and we can produce a simulated sample from the posterior density by sampling from a gamma density. This is conveniently done using the R function `rgamma`. A sample of 1000 draws from a gamma(331.6, 270.3) distribution is taken and stored in the vector `sim.post` by the R code

```
> sim.post = rgamma(1000, shape=331.6, rate=270.3)
```

One can visualize the posterior density by constructing a histogram or other type of density estimate on the sample of simulated draws. In the following code, we construct a histogram and overlay the exact gamma density (see Figure 12.3). The gamma density is a good match to the histogram, which indicates that the simulated sample is a good representation of the posterior.

```
> hist(sim.post, freq=FALSE, main="", xlab="Lambda")
> curve(dgamma(x, shape=331.6, rate=270.3), add=TRUE)
```

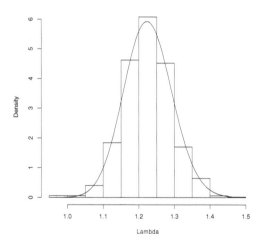

**Fig. 12.3** Histogram of a simulated sample from the posterior distribution of Madison's rate parameter. The exact gamma posterior is drawn on the graph, showing that the simulated sample seems to match the exact posterior.

One summarizes the posterior distribution of $\lambda$ using data analysis methods on the sample of simulated draws from the posterior probability distribution. The 5th, 50th and 95th quantiles of the posterior are found by computing the corresponding sample quantiles of the simulated sample of parameter values stored in `sim.post`.

```
> quantile(sim.post,c(0.05, 0.5, 0.95))
     5%       50%       95%
1.118563 1.225402 1.338694
```

The posterior median is given by the sample median 1.225 and a 90% interval estimate for $\lambda$ is bracketed by the 5th and 95th quantiles (1.119, 1.339).

**Comment:** One notes that the posterior median and interval estimate using the simulated draws are close to, but not exactly equal to the "exact" summary values computed using the `qgamma` function. Actually this will typically be the case. By simulating from the posterior distribution, we are introducing some errors due to the simulation procedure. So why do we simulate? There are a couple of good reasons. First, simulation provides an attractive way of learning about a probability distribution since we can use data analysis methods on the simulated sample to learn about the distribution. Second, we will shortly see that good methods exist for simulating from any posterior density (even those that are not convenient functional forms), so simulation provides a general all-purpose method for Bayesian computation.

## 12.6 Simulating a Probability Distribution by a Random Walk

### 12.6.1 Introduction

In our example, the posterior density turned out to be a convenient functional form (the gamma distribution) and one could use built-in R functions such as `dgamma` and `qgamma` to summarize the density. But in practice, the posterior density is not a familiar functional form and so one needs a general-purpose method for simulating from this distribution. In this section, we review one popular algorithm, the Metropolis-Hastings (M-H) random walk algorithm. In Chapter 13, we illustrate the use of this algorithm for simulating from a general sampling density and here we discuss the use of the same algorithm for simulating from an arbitrary probability distribution of one continuous parameter.

## 12.6.2 The Metropolis-Hastings random walk algorithm

Let $g(\theta)$ denote the posterior density of interest. One simulates a random walk through the probability distribution consisting of values $\theta^{(0)}, \theta^{(1)}, \theta^{(2)}, \ldots$ We specify this random walk by specifying a starting value $\theta^{(0)}$, and a rule for moving from the $t$th value in the sequence $\theta^{(t)}$ to the $(t+1)$st value in the sequence $\theta^{(t+1)}$.

a. Starting at $\theta^{(t)}$, simulate a candidate value $\theta^*$ uniformly from the interval $(\theta^{(t)} - C, \theta^{(t)} + C)$, where $C$ is an assigned constant determining the width of the proposal density.
b. Compute the posterior density at the current value $\theta^{(t)}$ and the candidate value $\theta^*$ – the values are $g(\theta^{(t)})$ and $g(\theta^*)$.
c. Compute the probability

$$PROB = \min\left(1, \frac{g(\theta^*)}{g(\theta^{(t)})}\right)$$

d. With probability $PROB$, the next value in the sequence is the candidate value $\theta^*$; otherwise the next value in the sequence is the current value $\theta^{(t)}$.

This random walk is simulating a special form of Markov Chain. Under general conditions, the sequence of simulated values $\theta^{(0)}, \theta^{(1)}, \theta^{(2)}, \ldots$ will converge to a sample from the posterior density $g(\theta)$. This algorithm is very similar to the random walk M-H algorithm described in Chapter 13. The only difference is that the algorithm in Chapter 13 uses a symmetric normal density to find the candidate and here we are using a uniform density.

In this algorithm, a candidate parameter value $\theta^*$ is randomly drawn from a neighborhood about the current value $\theta^{(t)}$. One needs to specify a value of the constant $C$ that determines the width of this neighborhood. As we will see in the example to follow, there is a range of suitable values of $C$ that will produce a simulated sample of draws that are approximately from the posterior density.

A short function $\mathtt{metrop.hasting.rw}$ is written to implement this random walk algorithm. (This function is very similar to the function of the same name in Chapter 13.) Before one uses this function, one has to first write a short function to compute the posterior density. For numerical accuracy, we compute the logarithm of the posterior density. In our example, the posterior density has a gamma form with shape 331.6 and rate= 270.3. The function $\mathtt{mylogpost}$ is written to compute the logarithm of this gamma density. The function has three arguments – the random parameter $\lambda$ and the shape and rate parameters. Note that we use the argument $\mathtt{log = TRUE}$ in the gamma density $\mathtt{dgamma}$ to compute the logarithm of the density.

```
mylogpost = function(lambda, shape, rate)
  dgamma(lambda, shape, rate, log=TRUE)
```

The function `metrop.hasting.rw` has five arguments: the name of the function `logpost` defining the logarithm of the posterior, the starting value `current` in the algorithm, the neighborhood width constant `C`, the simulation sample size `iter`, and any additional arguments needed in the function `logpost`. Here the additional arguments will be the shape and rate of the gamma density. Since this function is relatively short, it is instructive to explain the code of the function, line by line.

- Using the function `rep`, a vector `S` is defined with all 0's to store the simulated draws. A variable `n.accept` is defined that will be used to count the number of candidates that are accepted.

  ```
  S = rep(0, iter); n.accept = 0
  ```

- For each iteration in the loop,

  - we simulate a candidate $\theta^*$ in a neighborhood about the current value $\theta^{(t)}$
    ```
    candidate = runif(1, min=current - C, max=current + C)
    ```
  - we compute the probability of accepting the candidate
    ```
    prob = exp(logpost(candidate, ...) - logpost(current, ...))
    ```
  - we simulate a uniform random number – if the number is smaller than the probability, `accept="yes"`; otherwise `accept="no"`
    ```
    accept = ifelse(runif(1) < prob, "yes", "no")
    ```
  - if we accept, then the new current value is the candidate; otherwise the new current value remains the same value
    ```
    current = ifelse(accept == "yes", candidate, current)
    ```
  - we store the current value in the vector `S` and keep track of the number of accepted candidates
    ```
    S[j] = current; n.accept = n.accept + (accept == "yes")
    ```

- The function returns a list with two components: `S` is the vector of simulated draws and `accept.rate` is the proportion of accepted candidates.

  ```
  list(S=S, accept.rate=n.accept / iter)
  ```

The entire code for the function `metrop.hasting.rw` is displayed below.

```
metrop.hasting.rw = function(logpost, current, C, iter, ...){
  S = rep(0, iter); n.accept = 0
  for(j in 1:iter){
    candidate = runif(1, min=current - C, max=current + C)
    prob = exp(logpost(candidate, ...) - logpost(current, ...))
    accept = ifelse(runif(1) < prob, "yes", "no")
    current = ifelse(accept == "yes", candidate, current)
    S[j] = current; n.accept = n.accept + (accept == "yes")
  }
  list(S=S, accept.rate=n.accept / iter)
}
```

For our example, recall that the function `mylogpost` contained the definition of the logarithm of the gamma posterior density. Suppose we wish to start the random walk at $\lambda = 1$, simulate 1000 values of the Markov Chain, and use a scale value of $C = 0.2$. The final arguments to `metrop.hasting.rw` are the shape and rate values used in the `mylogpost` the function.

```
> sim = metrop.hastings.rw(mylogpost, 1, 0.2, 1000, 331.6, 270.3)
```

After one runs the random walk, some checks need to be done to see if the simulated draws are a "good" sample from the posterior density. In usual practice, one constructs a trace plot, a scatterplot of the sequence of simulated values $\theta^{(t)}$ against the iteration number $t$. If there is not a strong dependence pattern in the sequence and there is no general drift or movement across iterations, then the sequence will likely be a good simulated sample from the probability density of interest. Also one displays the acceptance rate, the rate at which candidates are accepted. Generally, acceptance rates between 20 and 40 percent are believed to produce acceptable simulated samples in the random walk M-H algorithm.

We illustrate the use of these checks for the algorithms for three different choices of the scale constant $C = 0.02, 0.2, 1.0$. Each of the random walks was run for 1000 iterations, each starting at the value $\lambda = 1$. Table 12.2 shows the acceptance rates for the three algorithms. If one uses a small neighborhood with $C = 0.02$, one is accepting practically all of the candidates; in contrast, with the scale value $C = 1.00$, only 10% of the candidates are accepted.

**Table 12.2** Acceptance rates for the random walk algorithm using three choices for the scale $C$.

| $C$ | Acceptance Rate |
|------|-----------------|
| 0.02 | 0.92 |
| 0.20 | 0.49 |
| 1.00 | 0.10 |

Figure 12.4 displays trace plots of the simulated samples for the three choices of $C$. Note that the choice $C = 0.02$ with a large acceptance rate of 92% displays a snake-like appearance. The simulated values are very correlated and one does not quickly explore the region where the density has most of its probability. In contrast, with the choice $C = 1.0$, the acceptance rate is 10% and it is common for the trace plot to have regions where the algorithm stays at the same value. To see which simulated sample is most accurate, Figure 12.5 displays density estimate plots of the simulated samples for the three choices of $C$. The dashed line in each graph is the exact gamma posterior density. The best simulated sample for representing the posterior density appears to be the choice $C = 0.2$, followed by $C = 1.0$ and $C = 0.02$.

The Metropolis-Hastings random walk algorithm is a good general-purpose algorithm for simulating from an arbitrary probability density. But it is not automatic in that one needs to experiment with several choices of $C$ and use acceptance rates and trace plots to decide on an appropriate choice of

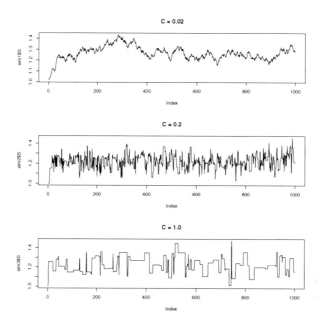

**Fig. 12.4** Trace plots of the simulated samples of the random walk algorithm using three choices of the scale constant $C$.

$C$. If one has an estimate of the standard deviation $\sigma$ of the density, then a reasonable starting guess at the scale is $C = 4\sigma$.

### 12.6.3 Using an alternative prior

One attractive aspect of the Bayesian approach is that there is great flexibility in the choice of models, and the random walk algorithm can be used to simulate from a general Bayesian model. To illustrate, suppose a second researcher has different beliefs about Madison's rate of using the word *from*. His best guess at $\lambda$ is 2 and he is 95% confident that $\lambda$ falls in the interval (1.6, 2.4). He represents this information by a normal prior with mean 2 and standard deviation 0.2. Using the "product of likelihood and prior" recipe, the posterior density is given (up to a proportionality constant) by

$$g(\lambda|\text{data}) \propto \exp(-262\lambda)\lambda^{323} \times \exp\left(-\frac{(\lambda-2)^2}{2(0.2)^2}\right).$$

This is not a conjugate problem as the posterior density is not a member of a convenient functional form like the gamma. But this is not a problem

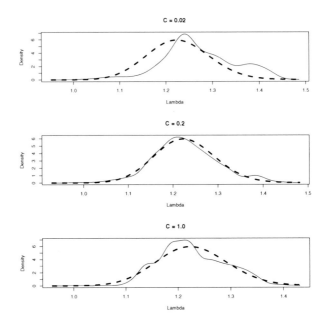

**Fig. 12.5** Density estimate plots of the simulated samples of the random walk algorithm using three choices of the scale constant $C$. The dashed lines in each graph correspond to the exact gamma density.

if we use our random walk simulation approach to summarize the posterior density. A function for computing the log posterior with the normal prior is displayed below. Since the likelihood is a gamma density with shape 324 and rate 262, we program the log likelihood using the `dgamma` function with the `log=TRUE` and the `dnorm` function with the `log=TRUE` option is used to program the log prior.

```
mylogpost = function(lambda){
 dgamma(lambda, shape=324, rate=262, log=TRUE) +
   dnorm(lambda, mean=2.0, sd=0.2, log=TRUE)
}
```

As an exercise, one is invited to use the function `metrop.hasting.rw` to simulate a sample from the posterior density.

## 12.7 Bayesian Model Checking

In our example, we have been assuming that the number of occurrences of *from* in a block of Madison's writing has a Poisson distribution. The validity of our inferences rests in our assumptions in the model that includes the assumption that we have Poisson sampling and that our beliefs about the rate parameter can be represented using the prior density $g(\lambda)$. But how do we check if the model assumptions are reasonable? Our model checking is based on the predictive distribution which is defined in the next section.

## 12.7.1 The predictive distribution

Suppose we have discovered a new sample of Madison's writings and we are interested in predicting the number of times the word *from* appears in this new sample. We denote this number as $y^*$ – we use the extra asterisk to refer to future data or data that has not already been observed. The values of $y^*$ are unknown and the probability distribution for this future data is called the *predictive* distribution.

Suppose we assume that, conditional on the rate $\lambda$, the future count $y^*$ has a Poisson distribution. Also assume that the current beliefs about the rate $\lambda$ are represented by a probability density $g(\lambda)$. Then one can simulate a value from the predictive density $f(y^*)$ by first simulating a rate from the density $g(\lambda)$ – call the simulated value $\lambda^*$. Then $y^*$ is simulated from the Poisson density $f(y^*|\lambda^*)$.

In our example, when we used a gamma prior, the posterior density for $\lambda$ was gamma(331.6, 270.3). Suppose we'd like to predict the number of *from*'s $y^*$ in a new block of writing. Using R code, we can simulate one value of $y^*$ using the `rgamma` function to simulate a value of $\lambda$ from its posterior and then using `rpois` and the simulated draw $\lambda^*$ to simulate a value of $y^*$.

```
> lambda = rgamma(1, shape=331.6, rate=270.3)
> ys = rpois(1, lambda)
```

This simulated draw is said to come from the *posterior* predictive distribution since we are using the posterior distribution $g(\lambda|\text{data})$ to model our current beliefs about the rate parameter.

Suppose we were interested in predicting the number of *from*'s in 262 new blocks of writing. We obtain a vector of word counts in this future sample by first simulating a single draw `lambda` from a gamma density and then simulating a sample of 262 draws from a Poisson distribution with this value of `lambda`.

```
> lambda = rgamma(1, shape=331.6, rate=270.3)
> ys = rpois(262, lambda)
```

We use the `table` function to tabulate the count values.

```
> table(ys)
ys
  0   1   2   3   4   5
 65 103  66  20   4   4
```

In 262 future blocks of writings, we predict that 65 of these blocks have none of this particular word, 103 blocks have exactly one word, 66 have two words, etc.

## 12.7.2 Model checking

When we predict the number of *from*'s in 262 blocks of writings, we notice that there are some differences between the counts in the observed sample and the counts in the predicted sample. For example, there were 90 blocks in our observed sample where we didn't find the word, and only 65 blocks in the predicted sample where we did not observe the word. Also in our sample, there were 3 blocks which had 6 *from*'s and there was no block in our predicted sample with more than 5 occurrences of the word. Of course, we just simulated a single predicted sample and compared with the observed data. But if simulated predictions from the model consistently look different from the observed counts, then this would make us question if our model is a good fit to the data. (Gelman et al. [19] give a general discussion of the use of the predictive distribution to check a Bayesian model.)

These observations motivate a method of checking our Bayesian model. We simulate a number of predicted samples from the posterior predictive distribution, where the size of each predictive sample is the same as the observed data. We compare these predictive samples with the observed data. If we see a pattern where each predictive dataset differs from the observed counts, then this suggests a lack-of-fit problem with our model.

We perform this comparison of observed with simulated data using the following R code. First a function `sim.y` is written to simulate and tabulate a sample of *from*'s in 262 writings from the posterior predictive distribution.

```
sim.y = function(){
  lambda = rgamma(1, shape=331.6, rate=270.3)
  ys = rpois(262, lambda)
  H = hist(ys, seq(-.5,9.5,1), plot=FALSE)
  data.frame(y=H$mids, f=H$counts)
}
```

Then we create a data frame containing the observed frequency counts and the simulated frequency counts from eight draws from the posterior predictive distributions. The `xyplot` function from the `lattice` package is used to display histograms of the values of $y^*$ from the actual data and these eight simulated samples. (See Figure 12.6.)

```
> D = data.frame(Type="Obs", y=c(0:6),
+    f = c(90, 93, 42, 17, 8, 9, 3))
> for(j in 1:8){
+    sim.out = sim.y()
+    D = rbind(D, data.frame(Type=paste("Sim",j),
+      y=sim.out$y, f=sim.out$f))
```

```
+ }
> library(lattice)
> xyplot(f ~ y | Type, data=D, type="h", lwd=3)
```

Generally, the histograms of the simulated distributions look similar to the observed data histogram. But if one focuses on the extreme values, one notes that is pretty uncommon for the simulated counts to be 4 or larger, and recall that there were 20 occurrences of 4 or more *from*'s in the observed blocks of text.

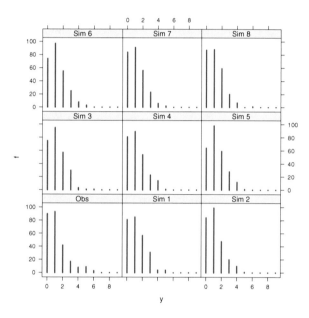

**Fig. 12.6** Histograms of eight samples from the posterior prediction distribution of the number of *from*'s in a future sample. The histogram of the observed data is shown in the lower left section of the plot.

Based on these comparisons between simulated predicted data and actual data, we think of a way of measuring the difference between the predicted and actual data distributions. Since the observed data seems to have more extreme values, we decide on focusing on the "checking function"

$$T = \text{ number of } y's \text{ where } y \geq 4.$$

For our data, the number of counts at least as large as 4 is

$$T_{obs} = 8 + 9 + 3 = 20.$$

Is it unusual to observe 20 blocks of writing with 4 or more *from*'s in predictions from our model? We answer this question by simulating 1000 samples from the posterior predictive distribution. For each sample, we compute the value of $T$, obtaining a sample of 1000 values of $T$ from its posterior predictive distribution. We construct a histogram of these simulated values and compare this distribution with the observed value of $T$, $T_{obs}$. If $T_{obs}$ is in the tail portion of this distribution, this indicates that the observed data is not consistent with predictive simulations from the model, and we should look for an alternative better-fitting model. Figure 12.7 displays a histogram of $T$ from our 1000 simulations. We see that a typical value of $T$ is about 9 and it is very likely that $T$ falls between 4 and 16. But, as the graph indicates, we observed a value of $T_{obs} = 20$ which is in the far right tail of the distribution.

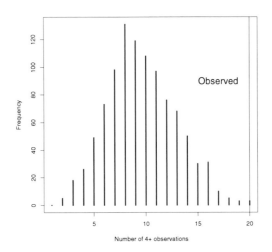

**Fig. 12.7** Posterior predictive distribution of blocks of "four or more *from*'s" in Madison's writings. The vertical line corresponds to the number of blocks of four or more *from*'s in the observed data.

Using this model checking method, we see that the observed counts of the specific word in Madison's blocks of writings has more variability that we would predict with this Poisson sampling model and the gamma prior. Since the likelihood function has the most impact on the posterior distribution, the main problem with our model appears to be the Poisson assumption and we need to look for an alternative sampling model that can accommodate this extra variability.

## 12.8 Negative Binomial Modeling

### 12.8.1 Overdispersion

Our basic assumption was that the number of occurrences $y$ of a specific word (like *from*) follows a Poisson distribution with mean $\lambda$. One property of the Poisson distribution is that the mean and variance are equal:

$$Var(y) = E(y).$$

In our example of counts in blocks of Madison's writing, one can estimate $V(y)$ and $E(y)$ by the sample variance and sample mean and one finds

$$s^2 = 1.83, \ \bar{y} = 1.23.$$

Here there is good evidence that the counts are *overdispersed* which means in this case that the variance exceeds the mean:

$$V(y) > E(y).$$

In this situation, the Poisson sampling assumption seems unreasonable and one wishes to fit an alternative sampling distribution that allows for overdispersion.

One distribution that allows for overdispersion is the negative binomial with probability function

$$p(y|a,b) = \frac{\Gamma(y+a)}{\Gamma(a)y!} \frac{b^a}{(b+1)^{a+y}}, \qquad y = 0,1,2,...$$

where $\Gamma(\cdot)$ is the complete gamma function. One can show that the mean and variance are given by

$$E(y) = \frac{a}{b}, \qquad Var(y) = \frac{a}{b}\left(1+\frac{1}{b}\right).$$

If we reparameterize this distribution in terms of the mean $\mu = a/b$ and $a$, then we can rewrite the probability function as

$$p(y|\mu,a) = \frac{\Gamma(y+a)}{\Gamma(a)y!} \left(\frac{a}{a+\mu}\right)^a \left(\frac{\mu}{a+\mu}\right)^y, \ y = 0,1,2,...$$

In this new parameterization, the mean and variance are

$$E(y) = \mu, \ Var(y) = \mu\left(1+\frac{\mu}{a}\right).$$

The parameter $a$ can be viewed as an overdispersion parameter since it determines how much larger the variance is than the mean. As $a$ is allowed to approach infinity, then the negative binomial model approaches the Poisson model where the mean and variance are equal.

## 12.8.2 Fitting the Negative Binomial model

It is more challenging to fit this negative binomial model than the Poisson since we have two unknown parameters, the mean parameter $\mu$ and the overdispersion parameter $a$. But we can use the same basic model-fitting approach as in the Poisson model. We construct a prior probability distribution for the parameters $(\mu, a)$, compute the likelihood function for the parameters, and then compute the posterior density by the "prior times likelihood" recipe.

### The prior

Suppose that our expert has little knowledge about the location of the parameters $(\mu, a)$ before any data is observed. In this case, one constructs a prior probability density on these two parameters which has a large spread that reflects imprecise prior beliefs. Here we assume $\mu$ and $a$ are independent, and each parameter is assigned a standard log-logistic density:

$$g(\mu, a) = \frac{1}{(1+\mu)^2} \frac{1}{(1+a)^2}, \qquad \mu > 0, \qquad a > 0.$$

For each parameter, the prior median is 1 and a 90% interval estimate is given by (0.11, 9.0). There are other possible choices for the prior that would reflect little information about the parameters, but this prior is sufficiently imprecise so that this prior information will have little impact on the posterior distribution.

### The likelihood

Recall that our observed data were the counts $y$ of the word *from* in many writings of James Madison. The likelihood is the probability of obtaining the observed data from Table 12.1 viewed as a function of the unknown parameters. If we assume negative binomial sampling, the likelihood is written as

$$L(\mu,a) = \prod_y p(y|\mu,a)$$

$$= \prod_y \frac{\Gamma(y+a)}{\Gamma(a)y!} \left(\frac{a}{a+\mu}\right)^a \left(\frac{\mu}{a+\mu}\right)^y.$$

Since the data is organized in the frequency table format where $y = 0$ is observed 90 times, $y = 1$ is observed 93 times, we can write the likelihood as

$$L(\mu,a) = p(0|\mu,a)^{90} p(1|\mu,a)^{93} p(2|\mu,a)^{42} p(3|\mu,a)^{17} p(4|\mu,a)^8$$
$$\times\, p(5|\mu,a)^9 p(6|\mu,a)^3,$$

where

$$p(y|\mu,a) = \frac{\Gamma(y+a)}{\Gamma(a)y!} \left(\frac{a}{a+\mu}\right)^a \left(\frac{\mu}{a+\mu}\right)^y.$$

**The posterior**

Assuming log-logistic priors for $(\mu,a)$ and a likelihood based on negative binomial sampling, the posterior is given, up to an unknown proportionality constant, by multiplying the likelihood by the prior:

$$g(\mu,a|\text{data}) \propto p(0|\mu,a)^{90} p(1|\mu,a)^{93} p(2|\mu,a)^{42} p(3|\mu,a)^{17} p(4|\mu,a)^8$$
$$\times\, p(5|\mu,a)^9 p(6|\mu,a)^3 \times \frac{1}{(1+\mu)^2} \frac{1}{(1+a)^2}.$$

To learn about the values of $(\mu,a)$, we apply the same simulation approach that we used for the Poisson sampling model. Using a random-walk algorithm, we simulate a large sample of simulated draws from the posterior probability distribution of the parameters, and then we summarize this simulated sample to perform inference.

To make the random walk algorithm more efficient, it helps if we transform the positive parameters $\mu$, $a$ to real-valued parameters using the log transformation:

$$\theta_1 = \log a, \quad \theta_2 = \log \mu.$$

The posterior probability distribution of the transformed parameters $(\theta_1, \theta_2)$ is given by

$$g(\theta_1,\theta_2|\text{data}) \propto p(0|\theta_1,\theta_2)^{90} p(1|\theta_1,\theta_2)^{93} p(2|\theta_1,\theta_2)^{42} p(3|\theta_1,\theta_2)^{17} p(4|\theta_1,\theta_2)^8$$
$$\times\, p(5|\theta_1,\theta_2)^9 p(6|\theta_1,\theta_2)^3 \times \frac{e^{\theta_1}}{(1+e^{\theta_1})^2} \frac{e^{\theta_2}}{(1+e^{\theta_2})^2},$$

where

$$p(y|\theta_1,\theta_2) = \frac{\Gamma(y+e^{\theta_1})}{\Gamma(e^{\theta_1})y!} \left(\frac{e^{\theta_1}}{e^{\theta_1}+e^{\theta_2}}\right)^a \left(\frac{e^{\theta_2}}{e^{\theta_1}+e^{\theta_2}}\right)^y.$$

If you look at the posterior density expression, you'll notice a new factor $\exp(\theta_1 + \theta_2)$ – this is a Jacobian term that is necessary when we apply a nonlinear transformation (such as $\log(a)$ and $\log(\mu)$) to the parameters $a$ and $\mu$.

## Random walking through the posterior

A straightforward generalization of the one-parameter Metropolis-Hastings random walk algorithm described earlier can be used to simulate from this two-parameter posterior density. As before, a starting value is chosen and the following plan is devised for moving from the $t$th value of the parameters, $\theta^{(t)} = (\theta_1^{(t)}, \theta_2^{(t)})$ to the $(t+1)$st value in the sequence. As before, we let $g(\theta)$ denote the posterior density where $\theta$ is the vector of parameters $(\theta_1, \theta_2)$.

a. A candidate value $\theta^* = (\theta_1^*, \theta_2^*)$ is simulated uniformly from the rectangle neighborhood $(\theta_1^{(t)} - C_1 \leq \theta_1^* \leq \theta_1^{(t)} + C_1, \theta_2^{(t)} - C_2 \leq \theta_2^* \leq \theta_2^{(t)} + C_2)$, where $C_1$ and $C_2$ are suitable constants controlling the size of the proposal region.
b. The posterior densities at the current value $\theta^{(t)} = (\theta_1^{(t)}, \theta_2^{(t)})$ and the candidate value $\theta^*$ are computed – the values are denoted by $g(\theta^{(t)})$ and $g(\theta^*)$.
c. The probability $P$ is computed, where

$$P = \min\left(1, \frac{g(\theta^*)}{g(\theta^{(t)})}\right).$$

d. With probability $P$, the next value in the sequence is the candidate value $\theta^*$; otherwise the next value in the sequence is the current value $\theta^{(t)}$.

The following R function `metrop.hasting.rw2` will implement the M-H random walk algorithm for a two-parameter probability density. The inputs are the function `logpost2` defining the logarithm of the posterior density, (2) the starting value at the parameter vector `curr`, the number of M-H iterations `iter`, and a vector of scale parameters `scale` consisting of the values of $C_1$ and $C_2$. The final ... inputs refers to any data or other inputs needed for the function `logpost2`. This function is very similar to the function `metrop.hasting.rw` used for the M-H algorithm in the one-parameter situation. Two lines are now used to simulate a candidate parameter value in the rectangle neighborhood of the current vector value. A matrix $S$ is used to store the vectors of simulated draws – the dimension of this matrix will be `iter` by 2 where each row corresponds to a simulated draw of $(\theta_1, \theta_2)$.

```
metrop.hasting.rw2 = function(logpost2, curr, iter, scale, ...){
  S = matrix(0, iter, 2)
  n.accept = 0; cand = c(0, 0)
  for(j in 1:iter){
    cand[1] = runif(1, min=curr[1] - scale[1],
```

```
    max=curr[1] + scale[1])
  cand[2] = runif(1, min=curr[2] - scale[2],
    max=curr[2] + scale[2])
  prob = exp(logpost2(cand, ...) - logpost2(curr, ...))
  accept = ifelse(runif(1) < prob, "yes", "no")
  if(accept == "yes") curr=cand
  S[j, ] = curr
  n.accept = n.accept + (accept == "yes")
  }
  list(S=S, accept.rate=n.accept / iter)
}
```

To use the function `metrop.hasting.rw2`, one writes a short function `lognbpost` to compute the logarithm of the posterior density. It is convenient to use the function `dnbinom` that computes the negative binomial probability function. Unfortunately, the negative binomial function in R uses a different parametrization – the function `dnbinom` is parameterized in terms of a size $n$ and a probability $p$ with probability function

$$f(y;n,p) = \frac{\Gamma(y+n)}{\Gamma(n)y!}p^n(1-p)^y, \; y = 0,1,2,...$$

Comparing with our representation, we see that $n = a = e^{\theta_1}$ and $p = a/(a+\mu) = e^{\theta_1}/(e^{\theta_1} + e^{\theta_2})$. In the function, the data is represented as a list consisting of y, a vector of the values of $y$, and f, a vector of the corresponding frequencies.

```
lognbpost = function(theta, data){
  sum(data$f * dnbinom(data$y, size=exp(theta[1]),
    prob=exp(theta[1]) / sum(exp(theta)), log=TRUE)) +
    sum(theta) - 2 * log(1 + exp(theta[1])) -
      2 * log(1 + exp(theta[2]))
}
```

To perform our simulation, we first define the data for our example.

```
> dat = list(y=0:6, f=c(90, 93, 42, 17, 8, 9, 3))
```

Then we run the function `metrop.hasting.rw2`, starting at value $(\theta_1, \theta_2) = (1,0)$, running the M-H algorithm for 10,000 iterations, and using the scale parameters $C_1 = 1.0$, $C_2 = 0.2$:

```
> sim.fit = metrop.hasting.rw2(lognbpost, c(1, 0), 10000, c(1.0, 0.2), dat)
```

The acceptance rate of this algorithm is 27%, which falls within the recommended acceptance rate in the literature for the Metropolis-Hastings random walk algorithm.

```
> sim.fit$accept.rate
[1] 0.2739
```

Based on the inspection of trace plots (not shown) of each parameter, we are satisfied that the simulated sample is a good approximation to the posterior. We display the joint posterior of $(\theta_1 = \log a, \theta_2 = \log \mu)$ by constructing a smoothed scatterplot (using the function `smoothScatter`) of the simulated pairs. (See Figure 12.8.) This scatterplot indicates that $\log a$ and $\log \mu$ are approximately independent parameters, and we can summarize the joint posterior by focusing on the marginal posterior distributions of each parameter.

```
> smoothScatter(sim.fit$S, xlab="log a", ylab="log mu")
```

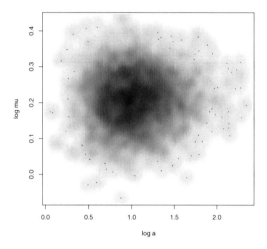

**Fig. 12.8** Smoothed scatterplot of simulated draws of $(\log a, \log \mu)$ from the posterior distribution in the negative binomial modeling of the word counts from Madison's writing.

### Inferences about the mean and overdispersion

Recall that we are fitting the negative binomial model since there appears to be overdispersion in the observed data. We measure the size of the overdispersion by means of the parameter $a$ – smaller values of $a$ indicate more overdispersion.

Our M-H simulation algorithm gives us a sample of values $\{(\theta_1^{(s)}, \theta_2^{(s)})\}$ from the posterior distribution of $\theta_1 = \log a$ and $\theta_2 = \log \mu$. The log transformations were used to facilitate the efficiency of the random walk algorithm, but we are more interested in learning the location of the overdispersion parameter $a$ instead of $\log a$. The simulated sample $\{\theta_1^{(s)}\}$ is a sample from the marginal posterior density of $\log a$. If we take the exponential transformation of these values, we obtain a sample from the posterior density of $a$.

```
> a = exp(sim.fit$S[ ,1])
```

We display the posterior density of the overdispersion parameter $a$ by constructing a density estimate of the simulated values in the vector **a**. (See Figure 12.9.)

```
> plot(density(a), xlab="a", ylab="Density", main="")
```

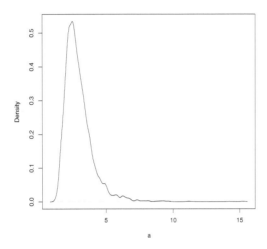

**Fig. 12.9** Density estimate of the marginal posterior of the overdispersion parameter $a$ for Madison writing example.

We see the overdispersion parameter $a$ has a right skewed distribution and most of the probability mass falls between 1 and 5. Using the `quantile` function, a 90% interval estimate for $a$ is constructed.

```
> quantile(a, c(0.05, 0.95))
      5%       95%
1.734907 4.853987
```

The overdispersion parameter falls in the interval $(1.73, 4.85)$ with probability 0.90.

In a similar fashion, we are interested in the average usage of the word **from** that corresponds to the mean parameter $\mu$. By taking the exponential of the simulated draws $\{\theta_2^{(s)}\}$ , we obtain a simulated sample of draws from the marginal posterior density of $\mu$ that we store in the vector **mu**.

```
> mu = exp(sim.fit$S[ ,2])
```

Figure 12.10 displays a density estimate of the marginal posterior of the mean usage from the simulated sample. A 90% Bayesian interval estimate for $\mu$ is again constructed by finding the corresponding sample quantiles from the simulated sample.

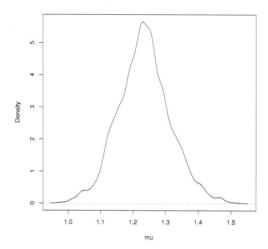

**Fig. 12.10** Density estimate of marginal posterior of the mean parameter $\mu$ for Madison writing example with negative binomial sampling.

```
> quantile(mu, c(0.05, 0.95))
      5%       95%
1.108965 1.371458
```

Comparing with the 90% interval estimate of $\mu$ obtained with Poisson sampling (1.119, 1.339), the negative binomial interval computed here is a bit wider. The longer interval is a consequence of the fact that, in negative binomial sampling, the overdispersion parameter $a$ is unknown and the uncertainty in this value results in a longer interval estimate. (For a similar reason, the interval estimate for a mean from a normal population with unknown variance will be longer than the corresponding interval for a normal mean when the variance is known.)

## Exercises

**12.1 (Learning about students sleep habits).** In Chapter 6, one is interested in the mean number of hours $\mu$ slept by students at a particular college. Suppose that a particular professor's prior beliefs about $\mu$ are represented by a normal curve with mean $\mu_0 = 6$ hours and variance $\tau_0^2 = 0.25$.

$$g(\mu) \propto \exp\left(-\frac{1}{2\tau_0^2}(\mu - \mu_0)^2\right).$$

a. Use the `curve` and `dnorm` functions to construct a plot of the prior density.

b. Use the `qnorm` function to find the quartiles of the prior density.

c. Find the probability (from the prior) that the mean amount of sleep exceeds 7 hours (Use the `pnorm` function.)

**12.2 (Learning about students sleep habits, continued).** Suppose the number of hours slept by a sample of $n$ students, $y_1, ..., y_n$, represent a random sample from a normal density with unknown mean $\mu$ and known variance $\sigma^2$. The likelihood function of $\mu$ is given by

$$L(\mu) = \exp\left(-\frac{n}{2\sigma^2}(\bar{y} - \mu)^2\right).$$

If we combine the normal prior density with this likelihood, it can be shown that the posterior density for $\mu$ also has the normal form with updated variance and mean parameters

$$\tau_1^2 = \left(\frac{1}{\tau_0^2} + \frac{n}{\sigma^2}\right)^{-1}, \mu_1 = \tau_1^2\left(\frac{\mu_0}{\tau_0^2} + \frac{n\bar{y}}{\sigma^2}\right).$$

a. For the sleeping data collected in Chapter 6, $n = 24$ students were sampled and the mean sleeping time was $\bar{y} = 7.688$. Assume that we know the sampling variance is given by $\sigma^2 = 2.0$. Use these values together with the prior mean and prior variance to compute the mean and variance of the posterior density of $\mu$.

b. Using two applications of the `curve` function, plot the prior and posterior densities together using contrasting colors.

c. Use the `qnorm` function to construct a 90% posterior interval estimate of the mean sleeping time.

d. Find the posterior probability that the mean amount of sleep exceeds 7 hours.

**12.3 (Waiting until a hit in baseball).** In sports, fans are fascinated with the patterns of streaky behavior of athletes. In baseball, a batter wishes to get a "base hit"; otherwise he records an "out." Suppose one records the number of outs between consecutive hits (the spacing) for a particular baseball player. For example, if the player begins the season with the outcomes H, O, O, H, H, O, O, O, O, H, O, H, H, then the spacings are given by 0, 2, 4, 1, 0 (one starts counting before the first outcome). One observes the following spacings for the player Ian Kinsler for the 2008 baseball season:

| | | | | | | | | | | | | | | | | | | |
|---|---|---|---|---|---|---|---|---|---|---|---|---|---|---|---|---|---|---|
| 0 | 2 | 0 | 4 | 1 | 0 | 2 | 0 | 1 | 0 | 0 | 1 | 1 | 3 | 1 | 0 | 0 | 0 | 1 |
| 6 | 0 | 9 | 0 | 4 | 1 | 9 | 1 | 0 | 3 | 4 | 5 | 5 | 1 | 0 | 2 | 4 | 0 | 4 |
| 0 | 3 | 2 | 1 | 0 | 1 | 3 | 7 | 0 | 3 | 1 | 2 | 14 | 4 | 0 | 1 | 6 | 1 | 10 |
| 1 | 2 | 0 | 1 | 0 | 4 | 5 | 0 | 7 | 3 | 1 | 2 | 1 | 2 | 1 | 2 | 2 | 4 | 3 |
| 3 | 1 | 1 | 2 | 1 | 2 | 7 | 0 | 3 | 1 | 2 | 2 | 2 | 2 | 0 | 3 | 4 | 1 | 1 |
| 0 | 0 | 1 | 1 | 1 | 11 | 2 | 2 | 1 | 3 | 1 | 0 | 1 | 2 | 1 | 1 | 1 | 0 | 0 |
| 2 | 0 | 10 | 1 | 2 | 2 | 1 | 1 | 3 | 1 | 1 | 0 | 0 | 1 | 0 | 1 | 0 | 1 | 1 |

| 0  | 1 | 0 | 0 | 0 | 2 | 1 | 4 | 5 | 5 | 0 | 0 | 0 | 0 | 2 | 0 | 8 | 5 | 2 |
| 11 | 8 | 0 | 7 | 1 | 3 | 1 |   |   |   |   |   |   |   |   |   |   |   |   |

Let $y$ denote the number of outs before the next base hit. A basic assumption is that $y$ has a geometric($p$) distribution with probability function

$$f(y|p) = p(1-p)^y, \ y = 0, 1, 2, \ldots$$

where $p$ is the player's hitting probability, If we assume independence in spacings, then the likelihood function of $p$ is given by

$$L(p) = \prod_y f(y|p) = p^n(1-p)^s,$$

where $n$ is the sample size and $s = \sum y$ is the sum of the spacings.

a. Compute the values of $n$ and $s$ for Kinsler's data.
b. Use the curve function to graph the likelihood function for values of $p$ between 0.2 and 0.5.
c. Based on the graph of the likelihood, which value of the hitting probability $p$ is "most likely" given the data?

**12.4 (Waiting until a hit, continued).** Based on Ian Kinsler's performance in previous seasons, a baseball fan has some prior beliefs about Kinsler's hitting probability $p$. She believes that $P(p < 0.300) = 0.5$ and $P(p < 0.350) = 0.90$. This prior information can be matched to a beta density with shape parameters $a = 44$ and $b = 102$.

$$g(p) = \frac{1}{B(44, 102)} p^{44-1}(1-p)^{102-1}, \ 0 < p < 1.$$

If one multiplies this prior density with the likelihood function found in Exercise 12.3, the posterior density for $p$ is given (up to a proportionality constant) by

$$g(p|\text{data}) \propto p^{n+44-1}(1-p)^{s+102-1}, \ 0 < p < 1,$$

where $n$ is the sample size and $s = \sum y$ is the sum of the spacings. This is a beta density with shape parameters $a = n + 44$ and $b = s + 102$. (The values of $n$ and $s$ are found from the data in Exercise 12.3.)

a. Using the curve function, graph the posterior density for values of the hitting probability $p$ between values 0.2 and 0.5.
b. Using the qbeta function, find the median of the posterior density of $p$.
c. Using the qbeta function, construct a 95% Bayesian interval estimate for $p$

**12.5 (Waiting until a hit, continued).** In Exercise 12.4, we saw that the posterior density for Ian Kinsler's hitting probability is a beta density with shape parameters $a = n + 44$ and $b = s + 102$. (The values of $n$ and $s$ are found from the data in Exercise 12.3.)

a. Using the `rbeta` function, simulate 1000 values from the posterior density of $p$.
b. Use the `hist` function on the simulated sample to display the posterior density.
c. Using the simulated draws, approximate the mean and standard deviation of the posterior density.
d. Using the simulated draws, construct a 95% Bayesian interval estimate. Compare the interval with the exact 95% interval estimate using the `qbeta` function.

**12.6 (Waiting until a hit, continued).** In Exercise 12.4, we saw that the posterior density for Ian Kinsler's hitting probability is a beta density with shape parameters $a = n + 44$ and $b = s + 102$. (The values of $n$ and $s$ are found from the data in Exercise 12.3.)

The function `metrop.hasting.rw` described in the chapter can be used to simulate a sample from the posterior density of the hitting probability $p$. The following function `betalogpost` will compute the logarithm of the beta density with shape parameters $a$ and $b$.

```
betalogpost = function(p, a, b)
  dbeta(p, a, b, log=TRUE)
```

a. Use the function `metrop.hasting.rw` together with the function `betalogpost` to simulate from the posterior density using the Metropolis Hastings random walk algorithm. Use $p = 0.2$ as a starting value, take 1000 iterations, and use the scale constant $C = 0.1$. Construct a trace plot of the simulated values and find the acceptance rate of the algorithm. Compute the posterior mean of $p$ from the simulated draws and compare the simulation estimate with the exact posterior mean $(n + 44)/(n + s + 44 + 102)$.
b. Rerun the random walk algorithm using the alternative scale constant values $C = 0.01$ and $C = 0.30$. In each case, construct a trace plot and compute the acceptance rate of the algorithm. Of the three choices for the scale constant $C$, are any of the values unsuitable? Explain.

# Chapter 13
# Monte Carlo Methods

## 13.1 The Monte Carlo Method of Computing Integrals

### 13.1.1 Introduction

A general problem in probability and statistical applications is the computation of an expectation of a random variable. Suppose the random variable $X$ has a density function $f(x)$ and we are interested in computing the expectation of a function $h(x)$ which is expressible as the integral

$$E(h(X)) = \int h(x)f(x)dx.$$

In the case where $h(x) = x$, this expectation corresponds to the mean $\mu = E(X)$. If $h(x) = (x - \mu)^2$, then the expectation corresponds to the variance $var(X) = E[(X - \mu)^2]$. If the function $h$ is the indicator function $h(x) = 1$ for $x \in A$, and $h(x) = 0$ elsewhere, then this expectation corresponds to a probability $E(h(X)) = P(A)$.

Suppose we are able to simulate a random sample of values $x^{(1)}, ..., x^{(m)}$ from the density $f(x)$. Then the Monte Carlo estimate of the expectation $E(h(X))$ is given by

$$\bar{h} = \frac{\sum_{j=1}^{m} h(x^{(j)})}{m}.$$

This is a good estimate of the expectation, since $\bar{h}$ will converge to $E(h(X))$ as the simulation sample size $m$ approaches infinity. But $\bar{h}$ is just a sample estimate of the expectation and so there will be some error in this simulation-based calculation. The variance of $\bar{h}$ is given by

$$Var(\bar{h}) = \frac{Var(h(X))}{m},$$

where $Var(h(X))$ is the variance of each simulated draw of $h(x^{(j)})$. We estimate $Var(h(X))$ by the sample variance of the simulated draws $\{h(x^{(j)})\}$:

$$\widehat{Var(h(X))} = \frac{\sum_{j=1}^{m}(h(x^{(j)}) - \bar{h})^2}{m-1}.$$

Then the associated standard error of this Monte Carlo estimate is given by

$$se_{\bar{h}} = \sqrt{\widehat{Var(\bar{h})}} \approx \sqrt{\frac{\widehat{Var(h(X))}}{m}}.$$

*Example 13.1 (Sleepless in Seattle).*
   Here is a simple example of the Monte Carlo method for computing integrals, inspired by the movie *Sleepless in Seattle*. Sam and Annie have a rendezvous at the Empire State Building on Valentine's Day. Sam arrives at a random time between 10 and 11:30 pm and Annie arrives at a random time between 10:30 and 12 am. What is the probability that Annie arrives before Sam? What is the expected difference in the two arrival times?

### 13.1.2 Estimating a probability

Let $A$ and $S$ denote respectively Annie's and Sam's arrival times, where we measure the arrival time as the number of hours after noon. We are assuming $A$ and $S$ are independent, where $A$ is uniformly distributed on $(10.5, 12)$ and $S$ is uniformly distributed on $(10, 11.5)$. We wish to compute the probability $P(A < S)$ which is expressed as the integral

$$P(A < S) = \int_{a<s} f_A(a)f_S(s)dads,$$

where $f_A$ and $f_S$ denote the uniform densities for $A$ and $S$. (By the independence assumption, the joint density $f_{A,S}(a,s) = f_A(a)f_S(s)$.) One can represent this probability as the expectation

$$P(A < S) = E[I(A < S)] = \int I(a < s)f_A(a)f_S(s)dads,$$

where $I(a < s)$ is the indicator function that is equal to one when $a < s$, and zero otherwise.
   To compute this integral by the Monte Carlo method, we simulate a large number of values, say 1000, from the distribution of $(A, B)$. Since $A$ and $B$ are independent, we perform this experiment by simulating 1000 values of $A$ from the uniform(10.5, 12) density and independently simulating 1000 values of $S$ from the uniform(10, 11.5) density. These simulations are conveniently

done using two applications of the `runif` function – the simulated draws are stored in the vectors `sam` and `annie`.

```
> sam = runif(1000, 10, 11.5)
> annie = runif(1000, 10.5, 12)
```

The probability $P(A < S)$ is estimated by the proportion of simulated pairs $(a, s)$ where $a$ is smaller than $s$. We count the number of pairs where $a < s$ by the `sum` function and divide this by the total number of simulations.

```
> prob = sum(annie < sam) / 1000
> prob
[1] 0.229
```

The estimated probability that Annie arrives before Sam is 0.229.

Figure 13.1 illustrates the process to compute this probability. The 1000 simulated pairs $\{S^{(j)}, A^{(j)}\}$ are displayed on a scatterplot and the shaded region represents the region where $A < S$. This display was generated using the R commands

```
> plot(sam, annie)
> polygon(c(10.5, 11.5, 11.5, 10.5),
+         c(10.5, 10.5, 11.5, 10.5), density=10, angle=135)
```

(The `polygon` function adds a shaded polygon to the current graph. The first two arguments are the vectors of horizontal and vertical coordinates of the polygon, and the `density` and `angle` arguments control the density and angle of the shading lines.) The Monte Carlo estimate of the probability is the proportion of points that fall in the shaded region.

Since the Monte Carlo estimate is simply a sample proportion, we can use the familiar formula for the standard error of a proportion to compute the standard error of this estimate. The standard error of a proportion estimate $\hat{p}$ is given by

$$se_{\hat{p}} = \sqrt{\frac{\hat{p}(1 - \hat{p})}{m}},$$

where $m$ is the simulation sample size. Applying this formula in R, we obtain

```
> sqrt(prob * (1 - prob) / 1000)
[1] 0.01328755
```

Applying a normal approximation for the sampling distribution of $\hat{p}$, we are 95% confident that the probability Annie will arrive earlier than Sam is between $0.229 - 1.96 \times 0.013$ and $0.229 + 1.96 \times 0.013$ or $(0.203, 0.255)$.

### 13.1.3 Estimating an expectation

Let's consider the second question: what is the expected difference in the arrival times? Since Annie is more likely to arrive later than Sam, we focus on computing the expectation $E(A - S)$ that can be expressed as the integral

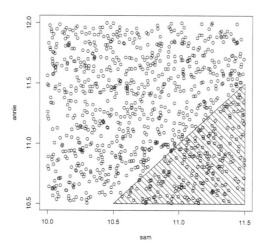

**Fig. 13.1** Scatterplot of simulated draws of $(S, A)$ for random meeting example. The shaded region corresponds to the event $\{A < S\}$ and the proportion of points in the region is an estimate of the probability $P(A < S)$.

$$E(A - S) = \int (a - s) f_A(a) f_S(s) da ds.$$

The Monte Carlo estimate of this expectation is given by

$$\widehat{E(A - S)} = \frac{\sum_{j=1}^{m} (A^{(j)} - S^{(j)})}{m},$$

where $\{(A^{(j)}, S^{(j)})\}$ is a simulated sample from the joint density of $(A, S)$. We already have simulated a sample from the joint density stored in the vectors `annie` and `sam`. For each pair of simulated values $(A^{(j)}, S^{(j)})$, we compute the difference $A^{(j)} - S^{(j)}$ and the differences are stored in the vector `difference`.

```
> difference = annie - sam
```

The Monte Carlo estimate of $E(A - S)$ is the mean of these differences. The estimated standard error of this sample mean estimate is the standard deviation of the differences divided by the square root of the simulation sample size.

```
> mc.est = mean(difference)
> se.est = sd(difference) / sqrt(1000)
> c(mc.est, se.est)
[1] 0.50156619 0.01945779
```

So we estimate that Annie will arrive 0.50 hours later than Sam; since the standard error is only 0.02 hours, we are 95% confident that the true difference is within 0.04 hours of this estimate.

## 13.2 Learning about the Sampling Distribution of a Statistic

*Example 13.2 (The sampling distribution of a median).*
    Suppose one observes a random sample $y_1, ..., y_n$ from the exponential density with unknown location parameter $\theta$.

$$f(y|\theta) = e^{-(y-\theta)}, \ y \geq \theta.$$

Suppose one only observes the sample median $M = \text{median}(y_1, ..., y_n)$; the actual observations $y_1, ..., y_n$ are not available. Using the available data, how can one construct a 90% confidence interval for $\theta$?
    A confidence interval for $\theta$ is found using the pivotal quantity method. Define a new random variable

$$W = M - \theta.$$

It can be shown that the distribution for $W$ does not depend on the unknown parameter. So it is possible to find two percentiles for $W$, $w_1$ and $w_2$, such that the probability that $(w_1, w_2)$ covers $W$ is a given probability $\gamma$:

$$P(w_1 < W < w_2) = \gamma.$$

Then it can be shown that a $100\gamma\%$ confidence interval for $\theta$ is given by $(M - w_2, M - w_1)$; the probability that this random interval covers $\theta$ is $\gamma$.

$$P(M - w_2 < \theta < M - w_1) = \gamma.$$

To implement this confidence interval method, one needs to obtain the sampling distribution of $W$ which represents the sample median of a random sample of size $n$ taken from the standard exponential density

$$f(w) = e^{-w}, \ w \geq 0.$$

Unfortunately, the sampling distribution of $W$ has a relatively messy form; the density in the special case where the sample size $n$ is odd is given by

$$f(m) = \frac{n!}{\left(\frac{n-1}{2}\right)! \left(\frac{n-1}{2}\right)!} e^{-m} \left(1 - e^{-m}\right)^{(n-1)/2} \left(e^{-m}\right)^{(n-1)/2}, \ m \geq 0.$$

Since the density does not have a convenient functional form, it is difficult to find the percentiles $w_1$ and $w_2$.

### 13.2.1 Simulating the sampling distribution by the Monte Carlo method

The distribution of the sample median $W$ is easily simulated by a Monte Carlo experiment. To obtain a single value of $W$, we use the `rexp` function to simulate a sample of size $n$ from a standard exponential distribution, and then use the `median` function to compute the sample median. We write a short function `sim.median` that will simulate a single value of $W$ given the sample size $n$.

```
> sim.median = function(n)
+    median(rexp(n))
```

To repeat this simulation a large number of times, it is convenient to use the `replicate` function. In the R code, we assume that we observe a sample of size $n = 21$ and simulate 10,000 values of the sample median $W$ that are stored in the vector M.

```
> M = replicate(10000, sim.median(21))
```

We display the simulated draws by a histogram computed using the `hist` function. (We use the `prob=TRUE` option so that the vertical axis corresponds to the probability density scale.)

```
> hist(M, prob=TRUE, main="")
```

To confirm that the simulated draws are truly from the sampling density of $M$, we first define a function `samp.med` to compute the exact density.

```
> samp.med = function(m, n){
+    con = factorial(n) / factorial((n - 1) / 2)^2
+    con * pexp(m)^((n - 1) / 2) * (1 - pexp(m))^((n - 1) / 2) * dexp(m)
+ }
```

Then we use the `curve` function to overlay the exact density on the histogram of the simulated sample. (See Figure 13.2.) Note that the curve is a good match to the histogram indicating that we have "captured" the density of interest using the Monte Carlo procedure.

```
> curve(samp.med(x, 21), add=TRUE, col="red")
```

### 13.2.2 Constructing a percentile confidence interval

Recall that we construct the confidence interval for $\theta$ by finding two values $w_1, w_2$ that bracket a specific probability $\gamma$ of the distribution of $W$. If we

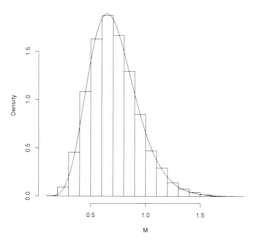

**Fig. 13.2** Histogram of the simulated draws from the sampling distribution of the sample median $M$ from an exponential distribution. The exact sampling density is plotted on top and it appears that the simulated draws are indeed distributed from the "correct" density.

desire a 90% confidence interval, then it is convenient to use the 5th and 95th percentiles. These percentiles are estimated by the 5th and 95th sample percentiles of the simulated sample of $W$ that we find using the `quantile` function.

```
> quantile(M, c(0.05, 0.95))
      5%       95%
0.398709 1.118365
```

Using these values, a 90% confidence interval for $\theta$ (for a sample of size 21) is given by $(M - 1.118365, M - 0.398709)$.

To illustrate using this confidence interval, suppose we collect the following 21 exponential observations that are stored in the vector y.

```
> y = c(6.11, 5.04, 4.83, 4.89, 4.97, 5.15, 7.98, 4.62,
+      6.19, 4.58, 6.34, 4.54, 6.37, 5.64, 4.53, 4.68,
+      5.17, 4.72, 5.06, 4.96, 4.70)
> median(y)
[1] 4.97
```

The sample median of this sample (found using the `median` function) is $M = 4.97$ and a 90% interval estimate for $\theta$ is given by:

```
> c(4.97 - 1.118365, 4.97 - 0.398709)
[1] 3.851635 4.571291
```

By the way, this data was simulated from an exponential distribution with value $\theta = 4.5$, so in this particular case, the confidence interval was successful in capturing the parameter value. If we repeated this procedure a large number of times, we would find that approximately 90% of the intervals would cover $\theta$.

**R$_{\mathbf{x}}$ 13.1** *It should be noted that this method for constructing a confidence interval is sample size dependent. If we had observed an exponential sample for a different value of $n$, then we would have to use a Monte Carlo experiment to simulate the sampling distribution of $M$ for that sample size.*

**R$_{\mathbf{x}}$ 13.2** *The endpoints of the confidence interval correspond to simulation-based estimates of the 5th and 95th percentiles of the distribution of $W$, and there will errors in these estimates. One can use a bootstrap procedure to obtain standard errors for these percentile estimates.*

## 13.3 Comparing Estimators

*Example 13.3 (The taxi problem).*
   A person is wandering the streets of a city and notices the following numbers of five taxis that pass by:

$$34, 100, 65, 81, 120.$$

Can she make an intelligent guess at the number of taxis in the city? This describes a simple problem in statistical inference where the population of interest is the collection of taxis that are driven in this city and one wishes to learn about the unknown number of taxis $N$.
   We assume that the taxis are numbered 1 through $N$ and the observer is equally likely to observe any one of the $N$ taxis at a given time. So one observes independent observations $y_1, ..., y_n$, where each observation is distributed according to the discrete uniform distribution

$$f(y) = \frac{1}{N}, \qquad y = 1, ..., N.$$

Consider two possible estimates of $N$: the largest taxi number observed

$$\hat{N}_1 = \max\{y_1, ..., y_n\}.$$

and twice the sample mean

$$\hat{N}_2 = 2\bar{y}.$$

Which is a better estimate of the number of taxis $N$?

## 13.3.1 A simulation experiment

We compare these two estimators using a simulation experiment. One simulates taxi numbers from a uniform distribution with a known number of taxis $N$ and computes the two estimates. By repeating this simulation process many times, one obtains two empirical sampling distributions. The two estimators can be compared by examining various properties of their respective sampling distributions.

The function `taxi` is written to implement a single simulation experiment. The function has two arguments, the actual number of taxis $N$ and the sample size $n$. The `sample` function simulates the observed taxi numbers and the values of the two estimates are stored in the variables `estimate1` and `estimate2`. The function returns the values of the two estimates.

```
> taxi = function(N, n){
+   y = sample(N, size=n, replace=TRUE)
+   estimate1 = max(y)
+   estimate2 = 2 * mean(y)
+   c(estimate1=estimate1, estimate2=estimate2)
+}
```

To illustrate this function, suppose the actual number of taxis in the city is $N = 100$ and we observe the numbers of $n = 5$ taxis.

```
> taxi(100, 5)
estimate1 estimate2
     93.0      87.6
```

Here the values of the two estimates are $N_1 = 93.0$ and $N_2 = 87.6$.

One can simulate this sampling process many times using the `replicate` function.

```
> EST = replicate(1000, taxi(100, 5))
```

The output variable `EST` is a matrix of 2 rows and 1000 columns where the two rows correspond to the simulated draws of the estimates $\hat{N}_1$ and $\hat{N}_2$ for the 1000 simulated experiments.

$R_x$ **13.3** *When one uses* `replicate` *with a function with multiple outputs, then the result is a matrix where the rows correspond to outputs and the columns correspond to replications. By default, R fills a matrix "by column," so the first set of outputs will be stored in the first column, and so on.*

## 13.3.2 Estimating bias

One desirable property of an estimator is unbiasedness which means that the average value of the estimator will be equal to the parameter in repeated sampling. In this setting, an estimator $\hat{N}$ is unbiased if $E(\hat{N}) = N$. When an estimator is biased, one wishes to estimate the bias

$$Bias = E(\hat{N}) - N.$$

In the case where one knows the parameter $N$, then one can estimate the bias of an estimator $\hat{N}$ in our simulation experiment by finding the sample mean of the $m$ simulated draws and subtracting the true parameter value:

$$\widehat{Bias} = \frac{1}{m} \sum_{j=1}^{m} \hat{N}^{(j)} - N.$$

The standard error of this simulation estimate is estimated by the standard deviation of the simulated draws $\{N^{(j)}\}$ divided by the square root of the simulation sample size.

In our simulation experiment, the simulated draws of the estimators are stored in the matrix EST and the true value of the number of taxis is $N = 100$. In the following R code, we estimate the bias of the estimators together with the corresponding standard errors.

```
> c(mean(EST["estimate1", ]) - 100, sd(EST["estimate1", ]) / sqrt(1000))
[1] -16.1960000    0.4558941
> c(mean(EST["estimate2", ]) -100, sd(EST["estimate2", ]) / sqrt(1000))
[1] -0.4248000  0.8114855
```

One can gauge the true sizes of the biases by computing 95% confidence intervals. The interval estimate for the bias of $\hat{N}_1$ is $(-16.196 - 2 \times 0.456, -16.196 + 2 \times 0.456) = (-17.108, -15.284)$ and the interval estimate for the bias of $\hat{N}_2$ is $(-0.425 - 2 \times 0.811, -0.425 + 2 \times 0.811) = (-2.047, 1.197)$. We conclude that the estimator $\hat{N}_1$ has a negative bias indicating that this estimator consistently underestimates the true number of taxis. In contrast, since the confidence interval for the bias of $\hat{N}_2$ includes the value zero, we cannot conclude the estimator $\hat{N}_2$ has a negative or positive bias.

### 13.3.3 Estimating mean distance from the target

Although the estimator $\hat{N}_2$ is preferable to $\hat{N}_1$ from the viewpoint of bias, it is not clear that it is a better estimator. Since it is desirable for an estimator to be close to the target parameter, it may be better to compare the estimators with respect to the mean distance from the parameter $N$ (this is commonly called the mean absolute error):

$$D = E(|\hat{N} - N|).$$

The quantity $D$ can be estimated by the average distance of the simulated estimator values $\hat{N}^{(1)}, ..., \hat{N}^{(m)}$ from $N$.

$$\hat{D} = \frac{1}{m} \sum_{j=1}^{m} |\hat{N}^{(j)} - N|.$$

We compute the absolute errors of the simulated draws of the two estimators $\hat{N}_1$ and $\hat{N}_2$ and store the absolute errors in the matrix `absolute.error`. Using the `boxplot` function, we compare the absolute errors of the two estimators. (Note that we apply the transpose function to convert `absolute.error` to a matrix with two columns that is needed for the `boxplot` function.) From Figure 13.3 we see that $\hat{N}_1 = \max y$ tends to have smaller estimation errors than $\hat{N}_2 = 2\bar{y}$.

```
> absolute.error = abs(EST - 100)
> boxplot(t(absolute.error))
```

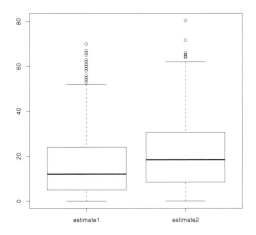

**Fig. 13.3** Comparison of the absolute errors in estimating the number of taxis using the two estimates $\hat{N}_1 = \max(y_i)$ and $\hat{N}_2 = 2\bar{y}$. It is clear that $\hat{N}_1$ tends to be closer than $\hat{N}_2$ to the target parameter value.

Using of the `apply` function to find the sample mean of the absolute errors, we estimate the mean distance $D$ of the two estimators.

```
> apply(t(absolute.error), 2, mean)
estimate1 estimate2
  16.7180    20.8172
```

A second application of the `apply` function is used to compute the simulation standard error of these estimates by the usual "standard deviation divided by the square root of the sample size" formula.

```
> apply(t(absolute.error), 2, sd) / sqrt(1000)
estimate1 estimate2
0.4500211 0.5093954
```

It is clear that the estimator $\hat{N}_1$ is preferable to $\hat{N}_2$ in that it has a smaller mean absolute error.


## 13.4 Assessing Probability of Coverage

*Example 13.4 (A popular confidence interval method for a proportion).*
To learn about the proportion $p$ of successes in a population, suppose one takes a random sample of size $n$ and computes the sample proportion $\hat{p}$. The popular Wald confidence interval for a proportion is based on the asymptotic normality of $\hat{p}$. For a given confidence level $\gamma$, this confidence interval is given by

$$INT_{Wald} = \left(\hat{p} - z\sqrt{\frac{\hat{p}(1-\hat{p})}{n}}, \hat{p} + z\sqrt{\frac{\hat{p}(1-\hat{p})}{n}}\right),$$

where $z$ denotes the corresponding $1 - (1-\gamma)/2$ percentile for a standard normal variable.

Since this procedure has confidence level $\gamma$, it is desirable that this interval contains the unknown proportion $p$ with coverage probability $\gamma$. That is, for any proportion value $p$ in the interval $(0, 1)$, we wish that

$$P(p \in INT_{Wald}) \geq \gamma.$$

Unfortunately, this procedure does not have the stated coverage probability. There are values of $p$ where the actual probability $P(p \in INT_{Wald})$ falls below the nominal level $\gamma$. This comment motivates several questions. Are there particular regions of the proportion space where the Wald interval has low coverage probability? Are there alternative confidence interval methods that are superior to the popular Wald interval from the perspective of coverage probability?


### 13.4.1 A Monte Carlo experiment to compute a coverage probability

To begin, we write a short function `wald` to compute the Wald interval. The inputs to the function are the number of successes `y`, the sample size `n`, and the confidence level **prob**. The output is a single-row matrix where the first column contains the lower endpoint and the second column contains the upper endpoint of the Wald interval. Looking at the code of the function, note that we use the **qnorm** function to compute the $z$ value and define **p** to be the proportion of successes.

```
wald = function(y ,n, prob){
  p = y / n
  z = qnorm(1 - (1 - prob) / 2)
  lb = p - z * sqrt(p * (1 - p) / n)
  ub = p + z * sqrt(p * (1 - p) / n)
  cbind(lb, ub)
}
```

Suppose we take a sample of 20, observe 4 successes, and desire a 90% interval estimate for the proportion $p$. We enter in this function and type

```
> wald(5, 20, 0.95)
            lb        ub
[1,] 0.0602273 0.4397727
```

One nice feature of the function **wald** is that it can be used to find the Wald intervals for a vector of values for $y$. If the argument **y** is a vector, then the variables **lb** and **ub** will be vectors of lower endpoints and upper endpoints and the output of **wald** will be a matrix where the rows give the confidence interval limits corresponding to the values for **y**. For example, suppose we wish to compute 90% Wald intervals for the $y$ values 2, 4, 6, 8 with a sample size of $n = 20$.

```
> y = c(2, 4, 6, 8)
> wald(y, 20, 0.90)
             lb        ub
[1,] -0.01034014 0.2103401
[2,]  0.05287982 0.3471202
[3,]  0.13145266 0.4685473
[4,]  0.21981531 0.5801847
```

We'll see shortly why it is important to permit vector arguments for $y$ in the **wald** function.

We are now ready to write a function **mc.coverage** to compute the actual probability of coverage of the Wald interval by a Monte Carlo experiment for a particular proportion value $p$. The arguments are the value of the proportion **p**, the sample size **n**, the stated confidence **prob**, and the number of Monte Carlo iterations **iter**. This function consists of three lines. First, we use the **rbinom** function to simulate **iter** values of the binomial random variable $y$ — these simulated draws are stored in the vector **y**. We compute the Wald intervals for each of the simulated binomial draws and the intervals are stored in the matrix **c.interval**. Last, we compute the fraction of intervals that actually contain the proportion value **p**. This is done using logical operators – **p** is contained in the interval if the proportion is greater than the left endpoint **and** the proportion is smaller than the right endpoint.

```
mc.coverage = function(p, n, prob, iter=10000){
  y = rbinom(iter, n, p)
  c.interval = wald(y, n, prob)
  mean((c.interval[ ,1] < p) & (p < c.interval[ ,2]))
}
```

Note that the function uses `iter` = 10000 iterations by default. Suppose we wish to compute the coverage probability of a 90% Wald interval with a sample size of $n = 20$ when the true proportion value is $p = 0.15$. We enter the function `mc.coverage` into the R console. Then the Monte Carlo estimate of the actual coverage probability is computed to be

```
> mc.coverage(0.15, 20, 0.90)
[1] 0.8005
```

The standard error of the Monte Carlo estimate of the probability is estimated by

```
> sqrt(0.8005 * (1 - 0.8005) / 10000)
[1] 0.003996245
```

The estimate of the actual probability of coverage is 0.8005 with a standard error of 0.004. Clearly the probability of coverage is smaller than the nominal level of $\gamma = 0.90$ which is a cause for some concern.

Actually we wish to learn about the probability of coverage of the Wald interval for all values of the proportion in the unit interval. We can perform this computation by the use of the `sapply` function. We write a short function `many.mc.coverage` to compute the coverage probability for a vector of values of $p$.

```
many.mc.coverage = function(p.vector, n, prob)
  sapply(p.vector, mc.coverage, n, prob)
```

The only difference between functions `mc.coverage` and `many.mc.coverage` is that `many.mc.coverage` accepts a vector argument `p.vector`. The `sapply` function will apply the Monte Carlo computation for each element of `p.vector` and output a vector of estimated coverage probabilities.

Now that we have a function `many.mc.coverage` for computing the coverage probability for a vector argument, we can plot the coverage probability as a function of the proportion $p$ using the `curve` function. Below we display the coverage probability of the Wald interval between the values $p = 0.001$ and $p = 0.999$ for the sample size $n = 20$ and confidence level $\gamma = 0.90$. Since we are interested in comparing the actual coverage probability with the confidence level, we add a horizontal line at 0.90 using the `abline` function. (See Figure 13.4.) We see that the Wald interval is not truly a 90% confidence interval since the probability of coverage is not uniformly greater than 0.90 over the range of proportion values. We see from the figure that the coverage probability is less than 0.90 for a large interval of proportion values, and the probability is smallest for extreme proportion values near zero and one. In an exercise, the reader will use a Monte Carlo experiment to explore the coverage probability of the improved "plus-four" confidence interval procedure.

```
> curve(many.mc.coverage(x, 100, 0.90), from=0.001, to=0.999,
+    xlab="p", ylab="Coverage Probability",
+    main=paste("n=", 100, ", prob=", 0.90),
+    ylim=c(0.7, 1))
> abline(h=.9)
```

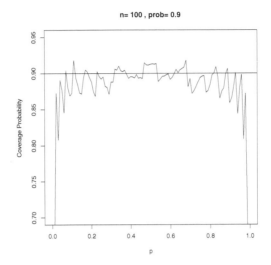

**Fig. 13.4** Estimated probability of coverage of the Wald interval procedure for the sample size $n = 20$ and confidence level $\gamma = 0.90$ using a Monte Carlo experiment with 10,000 iterations. The coverage probability is not uniformly over the stated value of 0.90, so this procedure is not a true 90% confidence interval.

## 13.5 Markov Chain Monte Carlo

### 13.5.1 Markov Chains

*Example 13.5 (A random walk).*

Suppose a person takes an unusual random walk on the values 1, 2, 3, 4, 5, 6 arranged in a circle – see Figure 13.5. If the person is currently at a particular location, in the next second she is equally likely to remain at the same location or move to an adjacent number. If she does move, she is equally likely to move left or right. This is a simple example of a discrete Markov chain. A Markov chain describes probabilistic movement between a number of states. Here there are six possible states, 1 through 6, corresponding to the possible locations of the walker. Given that the person is at a current location,

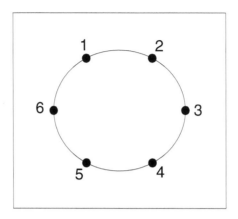

**Fig. 13.5** Six locations on a circle for the random walk example.

she moves to other locations with specified probabilities. The probability that she moves to another location depends only on her current location and not on previous locations visited. We describe movement between states in terms of transition probabilities – they describe the likelihoods of moving between all possible states in a single step in a Markov chain. We summarize the transition probabilities with a transition matrix $P$:

$$P = \begin{bmatrix} .50 & .25 & 0 & 0 & 0 & .25 \\ .25 & .50 & .25 & 0 & 0 & 0 \\ 0 & .25 & .50 & .25 & 0 & 0 \\ 0 & 0 & .25 & .50 & .25 & 0 \\ 0 & 0 & 0 & .25 & .50 & .25 \\ .25 & 0 & 0 & 0 & .25 & .50 \end{bmatrix}$$

The first row in $P$ gives the probabilities of moving to any states 1 through 6 in a single step from location 1, the second row gives the transition probabilities in a single step from location 2, and so on.

There are several important properties of this particular Markov chain. It is possible to go from every state to every state in one or more steps – a Markov chain with this property is said to be *irreducible*. Given that the person is in a particular state, if the person can only return to this state at regular intervals, then the Markov chain is said to be *periodic*. This example is *aperiodic* since it is not a periodic Markov chain.

We can represent one's current location as a probability row vector of the form

$$p = (p_1, p_2, p_3, p_4, p_5, p_6),$$

where $p_i$ represents the probability that the person is currently in state $i$. If $p^{(j)}$ represents the location of the traveler at step $j$, then the location of the traveler at the $j+1$ step is given by the matrix product

$$p^{(j+1)} = p^{(j)} P.$$

Suppose we can find a probability vector $w$ such that $wP = w$. Then $w$ is said to be the *stationary* distribution. If a finite Markov chain is irreducible and aperiodic, then it has a unique stationary distribution. Moreover, the limiting distribution of this Markov chain, as the number of steps approaches infinity, will be equal to this stationary distribution.

We can empirically demonstrate the existence of the stationary distribution of our Markov chain by running a simulation experiment. We start our random walk at a particular state, say location 3, and then simulate many steps of the Markov chain using the transition matrix $P$. The relative frequencies of our traveler in the six locations after many steps will eventually approach the stationary distribution $w$.

We first write a short R function to simulate a discrete Markov Chain. The arguments are the transition matrix P, the initial location of the chain starting.state, and steps, number of simulated steps through the chain. The output of the function is a vector that gives the locations of the chain in all of the steps. In the function, the function sample is used, together with the appropriate row of the transition matrix $P$, to do the sampling.

```
simulate.markov.chain = function(P, starting.state, steps){
  n.states = dim(P)[1]
  state = starting.state
  STATE = rep(0, steps)
  for(j in 1:steps){
    state = sample(n.states, size=1, prob=P[state, ])
    STATE[j ] = state
  }
  return(STATE)
}
```

We use the matrix function to input the transition matrix for our example.

```
> P = matrix(c(0.50, 0.25, 0, 0, 0,0.25,
+              0.25, 0.50, 0.25, 0, 0, 0,
+              0, 0.25, 0.50, 0.25, 0, 0,
+              0, 0, 0.25, 0.50, 0.25, 0,
+              0, 0, 0, 0.25, 0.50, 0.25,
+              0.25, 0, 0, 0, 0.25, 0.50),
+              nrow=6, ncol=6, byrow=TRUE)
> P
      [,1] [,2] [,3] [,4] [,5] [,6]
[1,]  0.50 0.25 0.00 0.00 0.00 0.25
[2,]  0.25 0.50 0.25 0.00 0.00 0.00
[3,]  0.00 0.25 0.50 0.25 0.00 0.00
[4,]  0.00 0.00 0.25 0.50 0.25 0.00
```

```
[5,] 0.00 0.00 0.00 0.25 0.50 0.25
[6,] 0.25 0.00 0.00 0.00 0.25 0.50
```

Then we simulate 10,000 steps through this Markov Chain by the function
simulate.markov.chain with the transition matrix P, the starting state 3,
and 10,000 steps. The vector s contains the locations of the chain in these
10,000 steps.

```
> s = simulate.markov.chain(P, 3, 10000)
```

The table function can be used to tabulate the locations of the chain in
the simulation.

```
> table(s)
s
   1    2    3    4    5    6
1662 1663 1653 1662 1655 1705
```

We can convert these counts to relative frequencies by dividing the output of
table by 10000.

```
> table(s) / 10000
     1      2      3      4      5      6
0.1662 0.1663 0.1653 0.1662 0.1655 0.1705
```

These relative frequencies approximately represent the stationary distribution
of this Markov Chain. It can be shown, using a separate calculation, that the
stationary distribution of this chain is uniform on the six states.

```
> w = c(1, 1, 1, 1, 1, 1) / 6
```

One can confirm that this vector is indeed the stationary distribution by
multiplying this vector by the transition matrix $P$ using the special matrix
multiplication operator.

```
> w %*% P
          [,1]      [,2]      [,3]      [,4]      [,5]      [,6]
[1,] 0.1666667 0.1666667 0.1666667 0.1666667 0.1666667 0.1666667
```

## 13.5.2 Metropolis-Hastings algorithm

A popular way of simulating from a general probability density $f(y)$ uses
the Markov chain Monte Carlo (MCMC) algorithm. This method essentially
is a continuous valued generalization of the discrete Markov chain setup described in the previous section. The MCMC sampling strategy sets up an irreducible, aperiodic Markov chain for which the stationary distribution equals
the probability distribution of interest. Here we introduce one general method
of constructing a Markov chain called the Metropolis-Hastings algorithm.

Suppose we wish to simulate from a particular density function $f(y)$.
We will simulate a Markov Chain and the simulated draws are denoted by
$y^{(1)}, y^{(2)}, y^{(3)}, \ldots$ A Metropolis-Hastings algorithm begins with an initial value

$y^{(0)}$ and a rule for simulating the next value in the sequence $y^{(t+1)}$ given the current value $y^{(t)}$. The rule for generating the next sequence value $x^*$ from the current value $x$ consists of three steps:

a. (**Proposal step**) Let $p(y|x)$ denote a proposal or candidate density of $y$ conditional on a value of $x$. If $x$ denotes the current value, then we simulate a value $y$ from the proposal density $p(y|x)$.

b. (**Acceptance probability**) The probability that this proposed value is accepted is given by the expression

$$PROB = \min\left( \frac{f(y)p(x|y)}{f(x)p(y|x)}, 1 \right).$$

c. (**Simulate the next value**) The next value in the sequence $x^*$ is simulated. It will be the proposed value $y$ with probability $PROB$; otherwise the next value $x^*$ in the sequence will be equal to the current value $x$.

Under some easily satisfied conditions on the proposal density $p(y|x)$, the sequence of simulated draws $y^{(1)}, y^{(2)}, y^{(3)}, \ldots$ will converge to a random variable that is distributed according to the probability density $f(y)$.

*Example 13.6 (The sampling distribution of a median (continued)).*
    Let's revisit the earlier example in this chapter where we were interested in the sampling distribution of the sample median $W$ from a random sample of size $n$ from a standard exponential density. One nice feature of the Metropolis-Hastings algorithm is that we don't need the normalizing constant of $f(y)$ since it will cancel out in the computation of the acceptance probability $PROB$. If we ignore the normalizing constant, we can write the sampling density of $W$ (again when the sample size is odd) as

$$f(y) \propto e^{-y} \left(1 - e^{-y}\right)^{(n-1)/2} \left(e^{-y}\right)^{(n-1)/2}, \ y \geq 0.$$

We begin by writing a short function to compute the logarithm of the density of interest $f(y)$. It is preferable to compute the log density instead of the density since the logarithm is more stable numerically (the computation will less likely result in an underflow result). The function has two arguments, the variable y and the sample size n, and the function uses the exponential density function **dexp** and the exponential cdf function **pexp**.

```
log.samp.med = function(y, n){
  (n - 1) / 2 * log(pexp(y)) + (n - 1) / 2 * log(1 - pexp(y)) +
    log(dexp(y))
}
```

Next, we need to think of a suitable choice for the proposal density $p(y|x)$. There are many suitable choices for $p$. Indeed one of the attractive features of this simulation algorithm is that it works for a large class of densities. Since the support of the density is positive values, we will try an exponential density – we let $p$ be an exponential density with rate parameter $x$:

$$p(y|x) = xe^{-xy}, \ y > 0.$$

With this choice of proposal density, the probability that a candidate value $y$ will be accepted (given that the current value is $x$) is given by

$$P = \min\left(\frac{f(y)ye^{-yx}}{f(x)xe^{-xy}}, 1\right).$$

Now we are ready to write a function `metrop.hasting.exp` to implement this Metropolis-Hastings algorithm using the exponential proposal density. There are four arguments to this function: `logf` is the name of the function that computes the logarithm of the density $f$ of interest, `current` is the initial value of the Markov Chain, `iter` is the number of steps of the chain, and ... refers to any parameters (such as $n$) that are needed in the function `logf`.

```
metrop.hasting.exp = function(logf, current, iter, ...){
  S = rep(0, iter); n.accept = 0
  for(j in 1:iter){
   candidate = rexp(1, current)
   prob = exp(logf(candidate, ...) - logf(current, ...) +
       dexp(current, candidate, log=TRUE) -
       dexp(candidate, current, log=TRUE))
   accept = ifelse(runif(1) < prob, "yes", "no")
   current = ifelse(accept == "yes", candidate, current)
   S[j] = current; n.accept = n.accept + (accept == "yes")
  }
  list(S=S, accept.rate=n.accept / iter)
}
```

Since this function is short, it is worthwhile to discuss the individual lines.

- Using the `rep` function, we allocate a vector of length `iter` that will store the simulated draws. Also we initialize the number of accepted draws `n.accept` to zero.
- We write a loop to repeat the basic Metropolis-Hastings step `iter` times. In each step,

  - We use the `rexp` function to simulate a candidate draw from an exponential distribution where the rate is the current value.
  - We compute the probability that the candidate will be accepted. We compute $\log f(x)$, $\log f(y)$, $\log p(y|x)$, $\log p(x|y)$ (using the `logf` and `dexp` functions), combine them and take the exponential, and the acceptance probability is stored in the variable `prob`.
  - We wish to accept the candidate with probability `prob`. We do this by simulating a uniform variable on the interval $(0, 1)$ (using the `runif` function), and accepting the candidate if the uniform variate is smaller than `prob`. The variable `accept` will either be "yes" or "no" depending on the result `runif(1) < prob`.
  - If we accept, the new current value is the candidate; otherwise, the new current value is the old current value.

- We store the new current value in the vector S and update the number of accepted draws

- The function returns a list with two components: S is the vector of simulated draws in the Markov Chain and accept.rate is the observed fraction of times the candidates were accepted.

Suppose we are interested in simulating the sampling distribution of the median for a given sample size $n = 21$, we plan on starting the Markov Chain at the value 1 and simulate 10,000 iterations of the Markov Chain. We perform this sampling by typing

```
> mcmc.sample = metrop.hasting.exp(log.samp.med, 1, 10000, 21)
```

By displaying the acceptance rate, we see that approximately 28% of the candidate draws were accepted in this algorithm

```
> mcmc.sample$accept.rate
[1] 0.2782
```

A MCMC simulation will, by theory, eventually converge to the density $f$ of interest. But this theoretical result says nothing about the speed of this convergence. In practice, one should be concerned about the length of the *burn-in* period, the number of iterations where the chain takes to reach the stationary distribution. Also, one would like the simulated sample to quickly explore the entire region where the density $f$ has most of its probability. If a MCMC simulation has this characteristic, we say that it has *good mixing*. In practice, it is helpful to construct a trace plot where one plots the simulated draws $\{y^{(t)}\}$ against the iteration number. In this example, we construct a trace plot of all simulated draws by typing

```
> plot(mcmc.sample$S)
```

The resulting display in shown in Figure 13.6.

In this example, we do not see any trend in the sequence of simulated draws, indicating that the length of the burn-in period is short. Also the simulated draws seem to have a wide range, indicating that the chain has mixed well and visited the "important" region of the density $f$.

Does this method work? Since we have the exact sampling density available, we can compare the sample of simulated draws with the exact density. The following R code displays a density estimate of the simulated draws and overlays the exact sampling density. (See Figure 13.7.) The two densities are very similar, indicating that this Monte Carlo algorithm has approximately converged to the density of interest.

```
> plot(density(mcmc.sample$S), lwd=2, main="", xlab="M")
> curve(samp.med(x, 21), add=TRUE, lwd=2, lty=2)
> legend("topright", c("Simulated", "Exact"), lty=c(1, 2),lwd=c(2, 2))
```

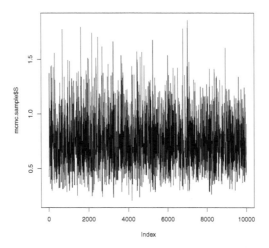

**Fig. 13.6** Index plot of the simulated draws from the Metropolis-Hastings algorithm for the median example with an exponential proposal density.

### 13.5.3 Random walk Metropolis-Hastings algorithm

Different Metropolis-Hastings algorithms are constructed depending on the choice of proposal density $p(y|x)$. One popular choice is to let $p$ have the form

$$p(y|x) = h(y - x),$$

where $h$ is a symmetric density about the origin. For example, suppose $h$ is the symmetric normal density with mean 0 and standard deviation $C$. Then the proposal density has the form

$$p(y|x) = \frac{1}{\sqrt{2\pi}C} \exp\left(-\frac{1}{2C^2}(y - x)^2\right).$$

In this case, the candidate value $y$ is chosen from a neighborhood of the current value $x$, where the width of the neighborhood is controlled by the normal standard deviation parameter $C$. If a "small" value of $C$ is chosen, then the candidate value will likely be close to the current value and the acceptance rate of the algorithm will be high. In contrast, a large value of $C$ will result in the candidate being chosen in a wide interval about the current value, lowering the acceptance rate of the method.

When the proposal density has this symmetric form, then $p(y|x)$ and $p(x|y)$ are equal and the acceptance probability has the simple form

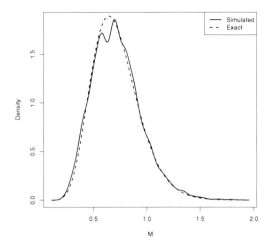

**Fig. 13.7** Density estimate of 10,000 simulated draws from the Metropolis-Hastings algorithm for the median example with an exponential proposal density. The exact sampling density is drawn with a dashed line.

$$PROB = \min\left(\frac{f(y)}{f(x)}, 1\right).$$

It is straightforward to write a function `metrop.hasting.rw` to implement the random walk Metropolis-Hastings algorithm with the normal proposal density by modifying a few lines of our earlier program. The only changes are that the candidate value is generated from a normal density with mean equal to the current value and standard deviation $C$, and there is a simplified form for the acceptance probability.

```
metrop.hasting.rw = function(logf, current, C, iter, ...){
  S = rep(0,  iter); n.accept = 0
  for(j in 1:iter){
    candidate = rnorm(1, mean=current, sd=C)
    prob = exp(logf(candidate, ...) - logf(current, ...))
    accept = ifelse(runif(1) < prob, "yes", "no")
    current = ifelse(accept == "yes", candidate, current)
    S[j] = current; n.accept = n.accept + (accept == "yes")
  }
list(S=S, accept.rate=n.accept / iter)
}
```

*Example 13.7 (The sampling distribution of a median (continued)).*
To use this random walk algorithm, one inputs a value of the scale parameter $C$. Below we simulate two MCMC samples for the median sampling problem, one with a value of $C = 1$ and the second with a value of $C = 0.05$.

```
> mcmc.sample1 = metrop.hasting.rw(log.samp.med, 1, 1, 10000, 21)
> mcmc.sample2 = metrop.hasting.rw(log.samp.med, 1, 0.05, 10000, 21)
```

Figure 13.8 displays traceplots of the simulated samples of the two samplers
and the titles of the plots display the corresponding acceptance rates.

```
> plot(mcmc.sample1$S, type="l", main=paste(
+     "Scale C = 1, Acceptance rate =",
+     round(mcmc.sample1$accept.rate,2)))
> plot(mcmc.sample2$S, type="l", main=paste(
+     "Scale C = 0.05, Acceptance rate =",
+     round(mcmc.sample2$accept.rate,2)))
```

When a small value of $C = 0.05$ is chosen, the acceptance rate is large (93%)
which results in simulated draws that are strongly positively correlated, and
the algorithm slowly explores the region of the target density. In contrast,
with the choice $C = 1$, the acceptance rate is small (26%), the simulated
values are less correlated, and the algorithm is more able to move across the
region where the density has large probability.

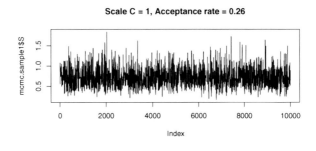

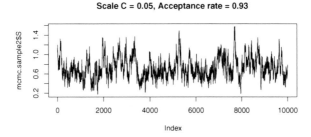

**Fig. 13.8** Trace plots of the simulated draws of the random walk Metropolis-Hastings
algorithm using two choices for the scale parameter $C$. Note that the choice $C = 0.05$
leads to a high acceptance rate, a snake-like appearance of the trace plot, and poorer
movement through the variable space.

## 13.5.4 Gibbs sampling

Gibbs sampling is an attractive method of designing a MCMC algorithm based on sampling from a set of conditional distributions. In the simplest setting, suppose one is interested in constructing a MCMC algorithm to simulate from a vector of variables $(x, y)$ distributed according to a joint density $g(x, y)$. Suppose it is convenient to sample from the conditional densities $g(x|y)$ and $g(y|x)$. Here "convenient" means that both of these conditional densities have familiar functional forms and can be simulated using standard simulation methods in R (such as `rnorm`, `rbeta`, `rgamma`, etc.).

Suppose that we start the MCMC algorithm at $x = x^{(0)}$. Then one cycle of the Gibbs sampler simulates

- $y^{(1)}$ from the conditional density $g(y|x^{(0)})$
- $x^{(1)}$ from the conditional density $g(x|y^{(1)})$

We continue sampling in this manner. We successively simulate from the two conditional densities $g(y|x)$ and $g(x|y)$, where in each case, we are conditioning on the most recently simulate draw of the other variable.

The Gibbs sampler defines a Markov chain and, under general conditions, the distribution of the variables after $t$ cycles $(x^{(t)}, y^{(t)})$, as $t$ approaches infinity, will converge to the joint distribution of interest $g(x, y)$. In many problems, the convergence will be pretty quick and so the sample of simulated draws $(x^{(1)}, y^{(1)}), ..., (x^{(m)}, y^{(m)})$ can be regarded as an approximate sample from the joint distribution of interest.

*Example 13.8 (Flipping a random coin).*

Suppose a hypothetical box contain a large amount of pennies. Each penny has a different chance of landing heads when flipped and the probabilities of heads follow a uniform density on the interval $(0, 1)$. You select a coin at random and flip it 10 times, obtaining 3 heads. You continue to flip the same coin 12 more times. What is the probability that you obtain exactly 4 heads?

In this problem, there are two unknown quantities – the probability $p$ that the coin lands heads, and the number of heads $y$ that we obtain in the final 12 flips. We can write down the joint density of $(p, y)$ from the description of the experiment – it is given by

$$g(p, y) = \binom{10}{3} p^3 (1-p)^7 \times \binom{12}{y} p^y (1-p)^{12-y}, \ 0 < p < 1, y = 0, 1, ..., 12.$$

To implement Gibbs sampling, one has to identify two conditional distributions, the density of $p$ conditional on $y$ and the density of $y$ conditional on $p$. To find $g(p|y)$ we collect all of the terms in the variable $p$ and regard $y$ as a constant – we find that

$$g(p|y) \propto p^{y+3} (1-p)^{19-y}, \qquad 0 < p < 1.$$

If we rewrite the density as

$$g(p|y) \propto p^{y+4-1}(1-p)^{20-y-1}, \qquad 0 < p < 1,$$

we recognize this conditional density as a beta density with shape parameters $a = y+4$ and $b = 20-y$. To get the second conditional density, we collect terms in $y$, regarding the probability $p$ as a constant. We obtain that

$$g(y|p) \propto \binom{12}{y} p^y (1-p)^{12-y}, \qquad y = 0, 1, ..., 12,$$

which we recognize as a binomial distribution with sample size 12 and probability of success $p$.

We write a short function **random.coin.gibbs** to implement the Gibbs sampling algorithm for this example. The function has two inputs – **p** is the starting value of the probability $p$ and **m** is the number of Gibbs sampling cycles. In the function, the matrix **S** is defined to store the simulated sample; for ease of identification, we label the two columns of the matrix using "p" and "y." A loop is set up; inside the loop, we successively sample $y$ from the binomial density (using the **rbinom** function) and $p$ from the beta density (using the **rbeta** function).

```
random.coin.gibbs = function(p=0.5, m=1000){
  S = matrix(0, m, 2)
  dimnames(S)[[2]] = c("p", "y")
  for(j in 1:m){
    y = rbinom(1, size=12, prob=p)
    p = rbeta(1, y+4, 20-y)
    S[j, ] = c(p, y)
  }
 return(S)
}
```

We run the Gibbs sampler with the default starting value $p = 0.5$ and 1000 cycles by typing **random.coin.gibbs** with no arguments.

```
> sim.values = random.coin.gibbs()
```

The variable **sim.values** contains the matrix of simulated draws from the joint distribution of $(p, y)$. One can draw a scatterplot of this distribution by the **plot** function. (See Figure 13.9.)

```
> plot(sim.values)
```

In this example, after inspection of trace plots of the simulated draws $\{p^{(t)}\}$ and $\{y^{(t)}\}$ (not shown), the burn-in period appears to be short and there is good mixing of the chain. Assuming the matrix **sim.matrix** is an approximate simulated sample from the joint probability distribution of $(p, y)$, the samples from the individual columns represent samples from the marginal distributions of $p$ and $y$. At the beginning of this section, we asked about the probability $P(y = 4)$. We tabulate the simulated values of $y$ by the **table** function.

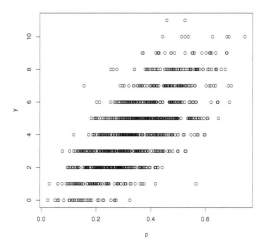

**Fig. 13.9** Scatterplot of simulated sample of $(p, y)$ from Gibbs sampling algorithm for random coin example.

```
> table(sim.values[ ,"y"])
  0   1   2   3   4   5   6   7   8   9  10  11
 47  92 135 148 168 152  85  81  55  22  11   4
```

In the sample of 1000 draws of $y$, we observed 168 fours, so $P(y = 4) \approx 168/1000 = 0.168$. Other properties of the marginal density of $y$, such as the mean and standard deviation, can be found by computing summaries of the sample of simulated draws of $y$.

## 13.6 Further Reading

Gentle [20] provides a general description of Monte Carlo methods. Chib and Greenberg [9] gives an introduction to the Metropolis-Hastings algorithm and Casella and George [8] give some basic illustrations of Gibbs sampling. Albert [1] provides illustrations of MCMC algorithms using R code.

## Exercises

**13.1 (Late to class?).** Suppose the travel times for a particular student from home to school are normally distributed with mean 20 minutes and standard deviation 4 minutes. Each day during a five-day school week she leaves home

30 minutes before class. For each of the following problems, write a short
Monte Carlo simulation function to compute the probability or expectation
of interest.

a. Find the expected total traveling time of the student to school for a five-
   day week. Find the simulation estimate and give the standard error for the
   simulation estimate.
b. Find the probability that the student is late for at least one class in the
   five-day week. Find the simulation estimate of the probability and the
   corresponding standard error.
c. On average, what will be the longest travel time to school during the five-
   day week? Again find the simulation estimate and the standard error.

**13.2 (Confidence interval for a normal mean based on sample quan-
tiles).** Suppose one obtains a normally distributed sample of size $n = 20$ but
only records values of the sample median $M$ and the first and third quartiles
$Q_1$ and $Q_3$.

a. Using a sample of size $n = 20$ from the standard normal distribution, simulate the sampling distribution of the statistic

$$S = \frac{M}{Q_3 - Q_1}.$$

   Store the simulated values of $S$ in a vector.
b. Find two values, $s_1, s_2$, that bracket the middle 90% probability of the
   distribution of $S$.
c. For a sample of size $n = 20$ from a normal distribution with mean $\mu$ and
   standard deviation $\sigma$, it can be shown that

$$P\left(s_1 < \frac{M - \mu}{Q_3 - Q_1} < s_2\right) = 0.90.$$

   Using this result, construct a 90% confidence interval for the mean $\mu$
d. In a sample of 20, we observe $(Q_1, M, Q_3) = (37.8, 51.3, 58.2)$. Using your
   work in parts (b) and (c), find a 90% confidence interval for the mean $\mu$.

**13.3 (Comparing variance estimators).** Suppose one is taking a sample
$y_1, ..., y_n$ from a normal distribution with mean $\mu$ and variance $\sigma^2$.

a. It is well known that the sample variance

$$S = \frac{\sum_{j=1}^{n}(y_j - \bar{y})^2}{n - 1}$$

   is an unbiased estimator of $\sigma^2$. To confirm this, assume $n = 5$ and perform
   a simulation experiment to compute the bias of the sample variance $S$.

b. Consider the alternative variance estimator

$$S_c = \frac{\sum_{j=1}^n (y_j - \bar{y})^2}{c},$$

where $c$ is a constant. Suppose one is interested in finding the estimator $S_c$ that makes the mean squared error

$$MSE = E\left[(S_c - \sigma^2)^2\right]$$

as small as possible. Again assume $n = 5$ and use a simulation experiment to compute the mean squared error of the estimators $S_3, S_5, S_7, S_9$ and find the choice of $c$ (among $\{3, 5, 7, 9\}$) that minimizes the MSE.

**13.4 (Evaluating the "plus four" confidence interval).** A modern method for a confidence interval for a proportion is the "plus-four" interval described in Agresti and Coull [2]. One first adds 4 imaginary observations to the data, two successes and two failures, and then apply the Wald interval to the adjusted sample. Let $\tilde{n} = n + 4$ denoted the adjusted sample size and $\tilde{p} = (y+2)/\tilde{n}$ denotes the adjusted sample proportion. Then the "plus-four" interval is given by

$$INT_{Plus-four} = \left(\tilde{p} - z\sqrt{\frac{\tilde{p}(1-\tilde{p})}{\tilde{n}}}, \hat{p} + z\sqrt{\frac{\tilde{p}(1-\tilde{p})}{\tilde{n}}}\right),$$

where $z$ denote the corresponding $1 - (1-\gamma)/2$ percentile for a standard normal variable.

By a Monte Carlo simulation, compute the probability of coverage of the plus-four interval for values of the proportion $p$ between 0.001 and 0.999. Contrast the probability of coverage of the plus-four interval with the Wald interval when the nominal coverage level is $\gamma = 0.90$. Does the plus-four interval have a 90% coverage probability for all values of $p$?

**13.5 (Metropolis-Hastings algorithm for the poly-Cauchy distribution).** Suppose that a random variable $y$ is distributed according to the poly-Cauchy density

$$g(y) = \prod_{i=1}^n \frac{1}{\pi(1 + (y - a_i)^2)},$$

where $a = (a_1, ..., a_n)$ is a vector of real-valued parameters. Suppose that $n = 6$ and $a = (1, 2, 2, 6, 7, 8)$.

a. Write a function to compute the log density of $y$. (It may be helpful to use the function `dcauchy` that computes the Cauchy density.)

b. Use the function `metrop.hasting.rw` to take a simulated sample of size 10,000 from the density of $y$. Experiment with different choices of the standard deviation $C$. Investigate the effect of the choice of $C$ on the acceptance rate, and the mixing of the chain over the probability density.

c. Using the simulated sample from a "good" choice of $C$, approximate the probability $P(6 < Y < 8)$.

**13.6 (Gibbs sampling for a Poisson/gamma model).** Suppose the vector of random variables $(X,Y)$ has the joint density function

$$f(x,y) = \frac{x^{a+y-1}e^{-(1+b)x}b^a}{y!\Gamma(a)}, \; x > 0, y = 0,1,2,...$$

and we wish to simulate from this joint density.

a. Show that the conditional density $f(x|y)$ has a gamma density and identify the shape and rate parameters of this density.
b. Show that the conditional density $f(y|x)$ has a Poisson density.
c. Write a R function to implement Gibbs sampling when the constants are given by $a = 1$ and $b = 1$.
d. Using your R function, run 1000 cycles of the Gibbs sampler and from the output, display (say, by a histogram) the marginal probability mass function of $Y$ and compute $E(Y)$.

# Appendix A
# Vectors, Matrices, and Lists

## A.1 Vectors

### *A.1.1 Creating a vector*

Vectors can be created in several ways. We have already seen two methods, the combine function c and the colon operator : for creating vectors with given values. For example, to create a vector of integers 1 through 9 and assign it to i, we can use either of the following.

```
i = c(1,2,3,4,5,6,7,8,9)
i = 1:9
```

To create a vector without specifying any of its elements, we only need to specify the type of vector and its length. For example,

```
y = numeric(10)
```

creates a numeric vector y of length 100, and

```
a = character(4)
```

creates a vector of 4 empty character strings. The vectors created are

```
> y
 [1] 0 0 0 0 0 0 0 0 0 0
> a
[1] "" "" "" ""
```

### *A.1.2 Sequences*

There are several functions in R that generate sequences of special types and patterns. In addition to :, there is a sequence function seq and a repeat rep function. We often use these functions to create vectors.

The **seq** function generates 'regular' sequences that are not necessarily integers. The basic syntax is **seq(from, to, by)**, where **by** is the size of the increment. Instead of **by**, we could specify the **length** of the sequence. To create a sequence of numbers $0, 0.1, 0.2, \ldots, 1$, which has 11 regularly spaced elements, either command below works.

```
> seq(0, 1, .1)
 [1] 0.0 0.1 0.2 0.3 0.4 0.5 0.6 0.7 0.8 0.9 1.0
> seq(0, 1, length=11)
 [1] 0.0 0.1 0.2 0.3 0.4 0.5 0.6 0.7 0.8 0.9 1.0
```

The **rep** function generates vectors by repeating a given pattern a given number of times. The **rep** function is easily explained by the results of a few examples:

```
> rep(1, 5)
[1] 1 1 1 1 1

> rep(1:2, 5)
 [1] 1 2 1 2 1 2 1 2 1 2

> rep(1:3, times=1:3)
[1] 1 2 2 3 3 3

> rep(1:2, each=2)
[1] 1 1 2 2

> rep(c("a", "b", "c"), 2)
[1] "a" "b" "c" "a" "b" "c"
```

## A.1.3 Extracting and replacing elements of vectors

If x is a vector, x[i] is the $i^{th}$ element of x. This syntax is used both for assignment and extracting values. If i happens to be a vector of positive integers $(i_1, \ldots, i_k)$, then x[i] is the vector containing $x_{i_1}, \ldots, x_{i_k}$, provided these are valid indices for x.

A few examples to illustrate extracting elements from a vector are:

```
> x = letters[1:8]    #letters of the alphabet a to h
> x[4]                 #fourth element
[1] "d"

> x[4:5]               #elements 4 to 5
[1] "d" "e"

> i = c(1, 5, 7)
> x[i]                 #elements 1, 5, 7
[1] "a" "e" "g"
```

Examples illustrating assignment are:

```
> x = seq(0, 20, 2)
> x[4] = NA              #assigns a missing value
> x
 [1]  0  2  4 NA  8 10 12 14 16 18 20

> x[4:5] = c(6, 7)      #assigns two values
> x
 [1]  0  2  4  6  7 10 12 14 16 18 20

> i = c(3, 5, 7)
> x[i] = 0        #assigns 3 zeros, at positions i
> x
 [1]  0  2  0  6  0 10  0 14 16 18 20
```

Sometimes it is easier to specify what to "leave out" rather than what to include. In this case, we can use negative indices. An expression x[-2], for example, is evaluated as all elements of x *except* the second one. The negative sign can also be used with a vector of indices, as shown below.

```
> x = seq(0, 20, 5)
> x
[1]  0  5 10 15 20

> i = c(1, 5)
> y = x[-i]              #leave out the first and last element
> y
[1]  5 10 15
```

## A.2 The sort and order functions

The sort function sorts the values of a vector in ascending order (by default) or descending order. At times, one wants to sort several variables according to the order of one of the variables. This can be accomplished with the order function.

*Example A.1 (The order function).* Suppose that we have the following data for temperatures and ozone levels

```
> temps = c(67, 72, 74, 62)
> ozone = c(41, 36, 12, 18)
```

and wish to sort the pairs (ozone, temps) in increasing order of ozone. The expression order(ozone) is a vector of indices that can be used to rearrange ozone into ascending order. This is the same order required for temps, so this order is what we need as an index vector for temps.

```
> oo = order(ozone)
> oo
[1] 3 4 2 1

> Ozone = sort(ozone)    #same as ozone[oo]
```

```
> Temps = temps[oo]

> Ozone
[1] 12 18 36 41
> Temps
[1] 74 62 72 67
```

**R<sub>x</sub> A.1** *In Example A.1 we used* **sort** *to sort the values of* **ozone**; *however, it is not really necessary to sort (again) because* **order(ozone)** *contains the information for sorting* **ozone**. *In this example,* **sort(ozone)** *is equivalent to* **ozone[oo]**.

See Section 2.7.3 for an example of sorting a data frame.

## A.3 Matrices

### A.3.1 Creating a matrix

A matrix can be created using the `matrix` function. The basic syntax is `matrix(x, nrow, ncol)`, where `x` is a constant or a vector of values, `nrow` is the number of rows and `ncol` is the number of columns. For example, to create a 4 by 4 matrix of 0's, we use

```
> matrix(0, 4, 4)
     [,1] [,2] [,3] [,4]
[1,]    0    0    0    0
[2,]    0    0    0    0
[3,]    0    0    0    0
[4,]    0    0    0    0
```

To create a matrix with specified elements, supply those elements in a vector as the first argument to `matrix`.

```
> X = matrix(1:16, 4, 4)
> X
     [,1] [,2] [,3] [,4]
[1,]    1    5    9   13
[2,]    2    6   10   14
[3,]    3    7   11   15
[4,]    4    8   12   16
```

The number of rows, number of columns, and dimension of a matrix that is already in the R workspace is returned by the functions `nrow` (or `NROW`), `ncol`, and `dim`, respectively.

```
> nrow(X)
[1] 4
> NROW(X)
[1] 4
> ncol(X)
```

```
[1] 4
> dim(X)
[1] 4 4
```

$R_X$ **A.2** (NROW, nrow, length) *It is helpful to know when to use NROW, nrow, or* length. length *gives the length of a vector, and* nrow *gives the number of rows of a matrix. But* length *applied to a matrix does not return the number of rows, but rather, the number of entries in the matrix. An advantage of NROW is that it computes* length *for vectors and* nrow *for matrices. This is helpful when the object could be either a vector or a matrix.*

Notice that when we supply the vector x=1:16 as the entries of the matrix X, the matrix was filled in column by column. We can change this pattern with the optional argument byrow=TRUE.

```
> matrix(1:16, 4, 4)
     [,1] [,2] [,3] [,4]
[1,]    1    5    9   13
[2,]    2    6   10   14
[3,]    3    7   11   15
[4,]    4    8   12   16

> A = matrix(1:16, 4, 4, byrow=TRUE)
> A
     [,1] [,2] [,3] [,4]
[1,]    1    2    3    4
[2,]    5    6    7    8
[3,]    9   10   11   12
[4,]   13   14   15   16
```

The row and column labels of the matrix also indicate how to extract each row or column from the matrix. To extract all of row 2 we use A[2,]. To extract all of column 4 we use A[,4]. To extract the element in row 2, column 4, we use A[2, 4].

```
> A[2,]
[1] 5 6 7 8
> A[,4]
[1]  4  8 12 16
> A[2,4]
[1] 8
```

A submatrix can be extracted by specifying a vector of row indices and/or a vector of column indices. For example, to extract the submatrix with the last two rows and columns of A we use

```
> A[3:4, 3:4]
     [,1] [,2]
[1,]   11   12
[2,]   15   16
```

To omit a few rows or columns we can use negative indices, in the same way that we did for vectors. To omit just the third row, we would use

```
> A[-3, ]
     [,1] [,2] [,3] [,4]
[1,]    1    2    3    4
[2,]    5    6    7    8
[3,]   13   14   15   16
```

The rows and columns of matrices can be named using the `rownames` or `colnames`. For example, we can create names for A as follows.

```
> rownames(A) = letters[1:4]
> colnames(A) = c("FR", "SO", "JR", "SR")
> A
  FR SO JR SR
a  1  2  3  4
b  5  6  7  8
c  9 10 11 12
d 13 14 15 16
```

Now one can optionally extract elements by name. To extract the column labeled "JR" we could use `A[, 3]` or

```
> A[, "JR"]
 a  b  c  d
 3  7 11 15
```

## A.3.2 Arithmetic on matrices

The basic arithmetic operators (`+` `-` `*` `/` `^`) on matrices apply the operations elementwise, analogous to vectorized operations. This means that if $A = (a_{ij})$ and $B = (b_{ij})$ are matrices with the same dimension, then `A*B` is a matrix of the products $a_{ij}b_{ij}$. Multiplying a matrix A above with itself using the `*` operator squares every element of the matrix.

```
> A = matrix(1:16, 4, 4, byrow=TRUE)
> A
     [,1] [,2] [,3] [,4]
[1,]    1    2    3    4
[2,]    5    6    7    8
[3,]    9   10   11   12
[4,]   13   14   15   16
> A * A
     [,1] [,2] [,3] [,4]
[1,]    1    4    9   16
[2,]   25   36   49   64
[3,]   81  100  121  144
[4,]  169  196  225  256
```

The exponentiation operator is also applied to each entry of a matrix. The R expression `A^2` is evaluated as the matrix of squared elements $a_{ij}^2$, not the matrix product $AA$.

```
> A^2        #not the matrix product
      [,1] [,2] [,3] [,4]
[1,]    1    4    9   16
[2,]   25   36   49   64
[3,]   81  100  121  144
[4,]  169  196  225  256
```

Matrix multiplication is obtained by the operator %*%. To obtain the square of matrix A (using matrix multiplication) we need A %*% A.

```
> A %*% A   #the matrix product
      [,1] [,2] [,3] [,4]
[1,]   90  100  110  120
[2,]  202  228  254  280
[3,]  314  356  398  440
[4,]  426  484  542  600
```

Many of the one-variable functions in R will be applied to individual elements of a matrix, also. For example, log(A) returns a matrix with the natural logarithm $log(a_{ij})$ as its entries.

```
> log(A)
          [,1]      [,2]     [,3]     [,4]
[1,] 0.000000 0.6931472 1.098612 1.386294
[2,] 1.609438 1.7917595 1.945910 2.079442
[3,] 2.197225 2.3025851 2.397895 2.484907
[4,] 2.564949 2.6390573 2.708050 2.772589
```

The apply function can be used to *apply* a function to rows or columns of a matrix. For example, we obtain the vector of column minimums and the vector of column maximums of A by

```
> apply(A, MARGIN=1, FUN="min")
[1]  1  5  9 13
```

```
> apply(A, MARGIN=2, FUN="max")
[1] 13 14 15 16
```

Row means and column means can be computed using apply or by

```
> rowMeans(A)
[1]  2.5  6.5 10.5 14.5
```

```
> colMeans(A)
[1]  7  8  9 10
```

The sweep function can be used to *sweep* out a statistic from a matrix. For example, to subtract the minimum of the matrix and divide the result by its maximum we can use

```
> m = min(A)
> A1 = sweep(A, MARGIN=1:2, STATS=m, FUN="-")   #subtract min

> M = max(A1)
> A2 = sweep(A1, 1:2, M, "/")  #divide by max
```

```
> A2
          [,1]        [,2]       [,3]       [,4]
[1,]  0.0000000  0.06666667  0.1333333  0.2000000
[2,]  0.2666667  0.33333333  0.4000000  0.4666667
[3,]  0.5333333  0.60000000  0.6666667  0.7333333
[4,]  0.8000000  0.86666667  0.9333333  1.0000000
```

Here we specified `MARGIN=1:2` indicating all entries of the matrix. The default function is subtraction, so the `"-"` argument could have been omitted in the first application of `sweep`.

The column mean can be subtracted from each column by

```
> sweep(A, 2, colMeans(A))
     [,1] [,2] [,3] [,4]
[1,]   -6   -6   -6   -6
[2,]   -2   -2   -2   -2
[3,]    2    2    2    2
[4,]    6    6    6    6
```

Example 1.8 in Section 1.3 illustrates matrix operations.

## A.4 Lists

One limitation of vectors, matrices, and arrays is that each of these types of objects may only contain one type of data. For example, a vector may contain all numeric data or all character data. A list is a special type of object that can contain data of multiple types.

Objects in a list can have names; if they do, they can be referenced using the `$` operator. Objects can also be referenced using double square brackets `[[ ]]` by name or by position.

*Example A.2 (Creating a list).* If the results of Example 1.3 (fatalities due to horsekicks) are not in the current workspace, run the script "horsekicks.R" discussed in Section 1.1.3. Now suppose that we would like to store the data (k and x), sample mean r and sample variance v, all in one object. The data are vectors x and k, both of length 5, but the mean and variance are each length 1. This data cannot be combined in a matrix, but it can be combined in a list. This type of list can be created as

```
> mylist = list(k=k, count=x, mean=r, var=v)
```

The contents of the list can be displayed like any other object, by typing the name of the object or by the `print` function.

```
> mylist
$k
[1] 0 1 2 3 4

$count
[1] 109  65  22   3   1
```

```
$mean
[1] 0.61

$var
[1] 0.6109548
```

Names of list components can be displayed by the **names** function.

```
> names(mylist)
[1] "k"     "count" "mean"  "var"
```

All of the components of this list have names, so they can be referenced by either name or position. Examples follow.

```
> mylist$count    #by name
[1] 109  65  22   3   1

> mylist[[2]]     #by position
[1] 109  65  22   3   1

> mylist["mean"] #by name
$mean
[1] 0.61
```

A compact way to describe the contents of a list is to use the **str** function. It is a general way to describe any object (**str** is an abbreviation for structure).

```
> str(mylist)
List of 4
 $ k    : num [1:5] 0 1 2 3 4
 $ count: num [1:5] 109 65 22 3 1
 $ mean : num 0.61
 $ var  : num 0.611
```

The **str** function is particularly convenient when the list is too large to read on one screen or the object is a large data set.

Many functions in R return values that are lists. An interesting example is the **hist** function, which displays a histogram; its return value (a list) is discussed in the next example.

*Example A.3 (The histogram object (a list)).* One of the many data sets included in the R installation is **faithful**. This data records waiting time between eruptions and the duration of the eruption for the Old Faithful geyser in Yellowstone National Park, Wyoming, USA. We are interested in the second variable **waiting**, the waiting time in minutes to the next eruption.

We construct a frequency histogram of the times between eruptions using the **hist** function, displayed in Figure A.1. The heights of the bars correspond to the counts for each of the bins. The histogram has an interesting shape. It is clearly not close to a normal distribution, and in fact it has two modes; one near 50-55, and the other near 80.

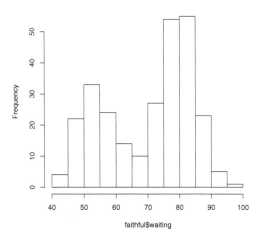

**Fig. A.1** Histogram of waiting times between Old Faithful eruptions in Example A.3.

Typically one is only interested in the graphical output of the **hist** function, but some useful information is returned by the function. That information can be saved if we assign the value of the function to an object. Here we saved the result of **hist** in an object H. The object H is actually a list; it stores a variety of information about the histogram such as the bin counts, breaks (endpoints), midpoints of intervals, etc.

```
> H = hist(faithful$waiting)
```

```
> names(H)
[1] "breaks"      "counts"      "intensities" "density"
[5] "mids"        "xname"       "equidist"
```

The endpoints of the intervals (the *breaks*) are

```
> H$breaks
 [1]  40  45  50  55  60  65  70  75  80  85  90  95 100
```

The frequencies in each interval (the bin *counts*) are

```
H$counts
 [1]  4 22 33 24 14 10 27 54 55 23  5  1
```

The help topic for **hist** describes each of the components of the list H in the *Value* section.

To create a list, use the **list** function; a simple example is at the beginning of this section (Example A.2). There are several other examples throughout this book that illustrate how to create lists.

# A.5 Sampling from a data frame

To draw a random sample of observations from a data frame, use the `sample` function to sample the row indices or the row labels, and extract these rows. Refer to the `USArrests` data introduced in Example 1.9 on page 20. To obtain a random sample of five states,

```
> i = sample(1:50, size=5, replace=FALSE)
> i
[1] 30  1  6 50  3
> USArrests[i, ]
           Murder Assault UrbanPop Rape
New Jersey    7.4     159       89 18.8
Alabama      13.2     236       58 21.2
Colorado      7.9     204       78 38.7
Wyoming       6.8     161       60 15.6
Arizona       8.1     294       80 31.0
```

Alternately, we could have sampled the row labels:

```
> samplerows = sample(rownames(USArrests), size=5, replace=FALSE)
> samplerows
[1] "North Dakota"  "West Virginia" "Montana"       "Idaho"
[5] "Virginia"
> USArrests[samplerows, ]
              Murder Assault UrbanPop Rape
North Dakota     0.8      45       44  7.3
West Virginia    5.7      81       39  9.3
Montana          6.0     109       53 16.4
Idaho            2.6     120       54 14.2
Virginia         8.5     156       63 20.7
```

# References

1. Albert, J. (2009), *Bayesian Computation with R*, Springer, New York.
2. Agresti, A. and Coull, B. (1998), "Approximate is better than 'exact' for interval estimation of binomial proportions," *The American Statistician*, 52, 119-126.
3. Ashenfelter, O. and Krueger, A. (1994), "Estimates of the economic return to schooling from a new sample of twins," *The American Economic Review*, 84, 1157–1173.
4. Bolstad, W. (2007), *Introduction to Bayesian Statistics*, Wiley-Interscience.
5. Box, G. E. P., Hunter W. G. and Hunter J. S. (1978), *Statistics for Experimenters: An Introduction to Design, Data Analysis, and Model Building*, Wiley, New York.
6. Cameron E. and Pauling L. (1978), "Experimental studies designed to evaluate the management of patients with incurable cancer," *Proc. Natl. Acad. Sci. U. S. A.*, 75, 4538–4542.
7. Canty, A. and Ripley. B. (2010), *boot: Bootstrap R Functions*, R package version 1.2-43.
8. Casella, G., and George, E. (1992), "Explaining the Gibbs sampler," *The American Statistician*, 46, 167–174.
9. Chib, S., and Greenberg, E. (1995), "Understanding the Metropolis-Hastings algorithm," *The American Statistician*, 49, 327–335.
10. Cleveland, W. (1979), "Robust locally weighted regression and smoothing scatterplots," *Journal of the American Statistical Association*, 74, 829–83.
11. Carmer, S. G. and Swanson, M. R. (1973), "Evaluation of ten pairwise multiple comparison proceedures by Monte Carlo methods," *Journal of the American Statistical Association*, 68, 66–74.
12. DASL: The Data and Story Library, `http://lib.stat.cmu.edu/DASL/`.
13. Davison, A. C. and Hinkley, D. V. (1997), *Bootstrap Methods and Their Applications*. Cambridge University Press, Cambridge.
14. Efron, B. and Tibshirani, R. J. (1993), *An Introduction to the Bootstrap*, Chapman & Hall/CRC, Boca Raton, FL.
15. Everitt, B. S., Landau, S., And Leese, M. (2001), *Cluster Analysis*, 4th edition, Oxford University Press, Inc. New York.
16. Faraway, J. (2002), "Practical regression and ANOVA Using R," Contributed documentation in PDF format available at `http://cran.r-project.org/doc/contrib/Faraway-PRA.pdf`.
17. Friendly, M. (2002), "A brief history of the mosaic display," *Journal of Computational and Graphical Statistics*, 11, 89-107.
18. Fienberg, S. E. (1971), "Randomization and social affairs: The 1970 draft lottery," *Science*, 171, 255–261.

19. Gelman, A., Carlin, J., Stern, H. and Rubin, D. (2003), *Bayesian Data Analysis*, second edition, Chapman and Hall, New York.
20. Gentle, J. E. (2003), *Random Number Generation and Monte Carlo Methods*, second edition, Springer, New York.
21. Hand D. J., Daly F., Lunn A. D., McConway K. J., Ostrowski E. (1994), *A Handbook of Small Data Sets*, Chapman and Hall, London.
22. Hoaglin, D., Mosteller, F., and Tukey, J. (2000), *Understanding Robust and Exploratory Data Analysis*, Wiley-Interscience.
23. Hoff, P. (2009), *A First Course in Bayesian Statistics*, Springer, New York.
24. Hogg, R. and Klugman, S. (1984), *Loss Distributions*, Wiley, New York.
25. Hollander. M. and Wolfe, D. (1999), *Nonparametric Statistical Methods*, second edition, Wiley, New York.
26. Hornik, K (2009). *R FAQ: Frequently Asked Questions on R*, Version 2.13.2011-04-07, `http://cran.r-project.org/doc/FAQ/R-FAQ.html`.
27. Hothorn, T., Hornik, K., van de Wiel, M. and Zeileis, A. (2008), "Implementing a class of permutation tests: The coin package," *Journal of Statistical Software*, 28, 1-23.
28. Koopmans, L. (1987), *Introduction to Contemporary Statistical Methods*, Duxbury Press.
29. Labby, Z. (2009), "Weldon's dice, automated" *Chance*, 22, 258-264.
30. Larsen, R. J. and Marx, M. L. (2006), *An Introduction to Mathematical Statistics and Its Applications*, 4th edition, Pearson Prentice Hall, Saddle River, New Jersey.
31. Leisch, F. (2007), *bootstrap: Functions for the Book "An Introduction to the Bootstrap,"* R package version 1.0-22.
32. Meyer, D., Zeileis, A., and Hornik, K. (2010), *vcd: Visualizing Categorical Data*, R package version 1.2-9.
33. Montgomery, D. G. (2001), *Design and Analysis of Experiments,* 5th edition., Wiley, New York.
34. Moore, G. (1965), "Cramming more components onto integrated circuits," *Electronics Magazine*, April, 114-117.
35. Mosteller, F. (2010), *The Pleasures of Statistics*, Springer, New York.
36. Mosteller, F. and Wallace, D. L. (1984), *Applied Bayesian and Classical Inference: The Case of the Federalist Papers*, Springer-Verlag, New York.
37. Murrell, P. (2006), *R Graphics*, Chapman and Hall, Boca Raton, Florida.
38. *OzDASL, Australian Data and Story Library*, `http://www.statsci.org/data/index.html`.
39. Pearson, K. (1900), "On the criterion that a given system of derivations from the probable in the case of a correlated system of variables is such that it can be reasonably supposed to have arisen from random sampling," *Philosophical Magazine*, 5, 157-175.
40. R Development Core Team (2011), "R: A language and environment for statistical computing," R Foundation for Statistical Computing, Vienna, Austria, `http://www.R-project.org/`.
41. The R Development Core Team (2011), "R: A Language and Environment for Statistical Computing, Reference Index." Version 2.13.0 (2011-04-13)
42. Ross, S. (2007), *Introduction to Probability Models*, 9th edition, Academic Press, Burlington, MA.
43. Sarkar, D. (2008), *Lattice: Multivariate Data Visualization with R*, Springer, New York.
44. Smith, J. M. and Misiak, H. (1973), "The effect of iris color on critical flicker frequency (CFF)," *Journal of General Psychology* 89, 91–95.
45. Snell, L. (1988), *Introduction to Probability*, Random House, New York.

46. Starr, N. (1997), "Nonrandom risk: the 1970 draft lottery. *Journal of Statistics Education*," 5, http://www.amstat.org/publications/jse/v5n2/datasets.starr.html

47. Tukey, J. (1977), *Exploratory Data Analysis*, Addison-Wesley.

48. "Campaign 2004: Time-tested formulas suggest both Bush and Kerry will win on Nov. 2," USA Today, http://www.usatoday.com/news/politicselections/nation/president/2004-06-23-bush-kerry-cover_x.htm.

49. Venables, W. N., Smith, D. M. and the R Development Core Team (2011). "An Introduction to R", Version 2.13.0 (2011-04-13).

50. Venables, W. N. and Ripley, B. D. (2002), *Modern Applied Statistics with S*, fourth edition. Springer, New York.

51. Verzani, J. (2010), *UsingR: Data sets for the text Using R for Introductory Statistics*," R package version 0.1-13.

52. Wickham, H. (2009), *ggplot2: Elegant Graphics for Data Analysis*, Springer, New York.

53. *The Washington Post* (2007) "Heads and Shoulders above the Rest,", http://blog.washingtonpost.com/44/2007/10/11/head_and_shoulders_above.html.

54. Wikipedia, "Heights of Presidents of the United States and presidential candidates," retrieved June 28, 2011.

55. Wilkinson, L. (2005), *The Grammar of Graphics*, second edition, Springer, New York.

56. Willerman, L., Schultz, R., Rutledge, J. N., and Bigler, E. (1991), "In vivo brain size and intelligence," *Intelligence*, 15, 223–228.

# Index